ADAPTIVE FILTERING PRIMER with MATLAB®

ADAPTIVE FILTERING PRIMER with MATLAB®

Alexander D. Poularikas
Zayed M. Ramadan

CRC
Taylor & Francis
Taylor & Francis Group
Boca Raton London New York

A CRC title, part of the Taylor & Francis imprint, a member of the
Taylor & Francis Group, the academic division of T&F Informa plc.

Published in 2006 by
CRC Press
Taylor & Francis Group
6000 Broken Sound Parkway NW, Suite 300
Boca Raton, FL 33487-2742

Printed in the United States of America on acid-free paper
10 9 8 7 6 5

International Standard Book Number-10: 0-8493-7043-4 (Softcover)
International Standard Book Number-13: 978-0-8493-7043-4 (Softcover)
Library of Congress Card Number 2005055996

Library of Congress Cataloging-in-Publication Data

Poularikas, Alexander D., 1933-
Adaptive filtering primer with MATLAB / by Alexander D. Poularikas and Zayed M. Ramadan.
p. cm.
Includes bibliographical references and index.
ISBN 0-8493-7043-4
1. Adaptive filters. 2. MATLAB. I. Ramadan, Zayed M. II. Title.

TK7872.F5P68 2006
621.3815'324--dc22 2005055996

informa
Taylor & Francis Group
is the Academic Division of Informa plc.

Visit the Taylor & Francis Web site at
http://www.taylorandfrancis.com

and the CRC Press Web site at
http://www.crcpress.com

Dedication

To my grandchildren Colton-Alexander and Thatcher-James, who have given us so much pleasure and happiness.

A.D.P.

To my great mom, Fatima, and lovely wife, Mayson, for their understanding, support, and love.

Z.M.R.

Preface

This book is written for the applied scientist and engineer who wants or needs to learn about a subject but is not an expert in the specific field. It is also written to accompany a first graduate course in digital signal processing. In this book we have selected the field of adaptive filtering, which is an important part of statistical signal processing. The adaptive filters have found use in many and diverse fields such as communications, control, radar, sonar, seismology, etc.

The aim of this book is to present an introduction to optimum filtering as well as to provide an introduction to realizations of linear adaptive filters with finite duration impulse response. Since the signals involved are random, an introduction to random variables and stochastic processes are also presented.

The book contains all the material necessary for the reader to study its contents. An appendix on matrix computations is also included at the end of the book to provide supporting material. The book includes a number of MATLAB® functions and m-files for practicing and verifying the material in the text. These programs are designated as Book MATLAB Functions. The book includes many computer experiments to illustrate the underlying theory and applications of the Wiener and adaptive filtering. Finally, at the end of each chapter (except the first introductory chapter) numerous problems are provided to help the reader develop a deeper understanding of the material presented. The problems range in difficulty from undemanding exercises to more elaborate problems. Detailed solutions or hints and suggestions for solving all of these problems are also provided.

Additional material is available from the CRC Web site, www.crcpress.com. Under the menu Electronic Products (located on the left side of the screen), click Downloads & Updates. A list of books in alphabetical order with Web downloads will appear. Locate this book by a search or scroll down to it. After clicking on the book title, a brief summary of the book will appear. Go to the bottom of this screen and click on the hyperlinked "Download" that is in a zip file.

MATLAB® is a registered trademark of The Math Works, Inc. and is used with permission. The Math Works does not warrant the accuracy of the text or exercises in this book. This book's use or discussion of MATLAB® software or related products does not constitute endorsement or sponsorship by The

For product information, please contact:

The Math Works, Inc.
3 Apple Hill Drive
Natick, MA 01760-2098 USA
Tel: 508-647-7000
Fax: 508-647-7001
E-mail: info@mathworks.com
Web: www.mathworks.com

Authors

Alexander D. Poularikas received his Ph.D. from the University of Arkansas and became professor at the University of Rhode Island. He became chairman of the Engineering Department at the University of Denver and then became chairman of the Electrical and Computer Engineering Department at the University of Alabama in Huntsville. He has published six books and has edited two. Dr. Poularikas served as editor-in-chief of the *Signal Processing* series (1993–1997) with ARTECH HOUSE and is now editor-in-chief of the *Electrical Engineering and Applied Signal Processing* series as well as the *Engineering and Science Primers* series (1998–present) with Taylor & Francis. He was a Fulbright scholar, is a lifelong senior member of IEEE, and is a member of Tau Beta Pi, Sigma Nu, and Sigma Pi. In 1990 and 1996, he received the Outstanding Educator Award of IEEE, Huntsville Section.

Zayed M. Ramadan received his B.S. and M.S. degrees in electrical engineering (EE) from Jordan University of Science and Technology in 1989 and 1992, respectively. He was a full-time lecturer at Applied Science University in Jordan from 1993 to 1999 and worked for the Saudi Telecommunications Company from 1999 to 2001. Dr. Ramadan enrolled in the Electrical and Computer Engineering Department at the University of Alabama in Huntsville in 2001, and received a second M.S. in 2004 and a Ph.D. in 2005, both in electrical engineering and both with honors. His main research interests are adaptive filtering and their applications, signal processing for communications, and statistical digital signal processing.

Contents

Chapter 1 Introduction **1**
1.1 Signal processing 1
1.2 An example 1
1.3 Outline of the text 2

Chapter 2 Discrete-time signal processing **5**
2.1 Discrete-time signals 5
2.2 Transform-domain representation of discrete-time signals 5
2.3 The Z-Transform 11
2.4 Discrete-time systems 13
Problems 17
Hints-solutions-suggestions 17

Chapter 3 Random variables, sequences, and stochastic processes **19**
3.1 Random signals and distributions 19
3.2 Averages 22
3.3 Stationary processes 26
3.4 Special random signals and probability density functions 29
3.5 Wiener–Khintchin relations 32
3.6 Filtering random processes 34
3.7 Special types of random processes 36
3.8 Nonparametric spectra estimation 40
3.9 Parametric methods of power spectral estimations 49
Problems 51
Hints-solutions-suggestions 52

Chapter 4 Wiener filters **55**
4.1 The mean-square error 55
4.2 The FIR Wiener filter 55
4.3 The Wiener solution 59
4.4 Wiener filtering examples 63
Problems 73
Hints-solutions-suggestions 74

Chapter 5 Eigenvalues of R_x — properties of the error surface 77
5.1 The eigenvalues of the correlation matrix 77
5.2 Geometrical properties of the error surface 79
Problems 81
Hints-solutions-suggestions 82

Chapter 6 Newton and steepest-descent method 85
6.1 One-dimensional gradient search method 85
6.2 Steepest-descent algorithm 91
Problems 96
Hints-solutions-suggestions 97

Chapter 7 The least mean-square (LMS) algorithm 101
7.1 Introduction 101
7.2 Derivation of the LMS algorithm 102
7.3 Examples using the LMS algorithm 104
7.4 Performance analysis of the LMS algorithm 112
7.5 Complex representation of LMS algorithm 126
Problems 129
Hints-solutions-suggestions 130

Chapter 8 Variations of LMS algorithms 137
8.1 The sign algorithms 137
8.2 Normalized LMS (NLMS) algorithm 139
8.3 Variable step-size LMS (VSLMS) algorithm 141
8.4 The leaky LMS algorithm 142
8.5 Linearly constrained LMS algorithm 145
8.6 Self-correcting adaptive filtering (SCAF) 150
8.7 Transform domain adaptive LMS filtering 153
8.8 Error normalized LMS algorithms 158
Problems 167
Hints-solutions-suggestions 167

Chapter 9 Least squares and recursive least-squares signal processing 171
9.1 Introduction to least squares 171
9.2 Least-square formulation 171
9.3 Least-squares approach 180
9.4 Orthogonality principle 182
9.5 Projection operator 184
9.6 Least-squares finite impulse response filter 186
9.7 Introduction to RLS algorithm 188
Problems 197
Hints-solutions-suggestions 197

Abbreviations **203**

Bibliography **205**

Appendix — Matrix analysis **207**
A.1 Definitions 207
A.2 Special matrices 210
A.3 Matrix operation and formulas 212
A.4 Eigen decomposition of matrices 215
A.5 Matrix expectations 217
A.6 Differentiation of a scalar function with respect to a vector 217

Index **219**

chapter 1

Introduction

1.1 *Signal processing*

In numerous applications of signal processing and communications we are faced with the necessity to remove noise and distortion from the signals. These phenomena are due to time-varying physical processes, which sometimes are unknown. One of these situations is during the transmission of a signal (message) from one point to another. The medium (wires, fibers, microwave beam, etc.), which is known as the *channel*, introduces noise and distortion due to the variations of its properties. These variations may be slow varying or fast varying. Since most of the time the variations are unknown, it is the use of *adaptive* filtering that diminishes and sometimes completely eliminates the signal distortion.

The most common adaptive filters, which are used during the adaptation process, are the finite impulse response filters (FIR) types. These are preferable because they are stable, and no special adjustments are needed for their implementation.

The adaptation approaches, which we will introduce in this book, are: the Wiener approach, the *least-mean-square* algorithm (LMS), and the *least-squares* (LS) approach.

1.2 *An example*

One of the problems that arises in several applications is the identification of a system or, equivalently, finding its input-output response relationship. To succeed in determining the filter coefficients that represent a model of the unknown system, we set a system configuration as shown in Figure 1.2.1.

The input signal, $\{x(n)\}$, to the unknown system is the same as the one entering the adaptive filter. The output of the unknown system is the desired signal, $\{d(n)\}$. From the analysis of linear time-invariant systems (LTI), we know that the output of linear time-invariant systems is the convolution of their input and their impulse response.

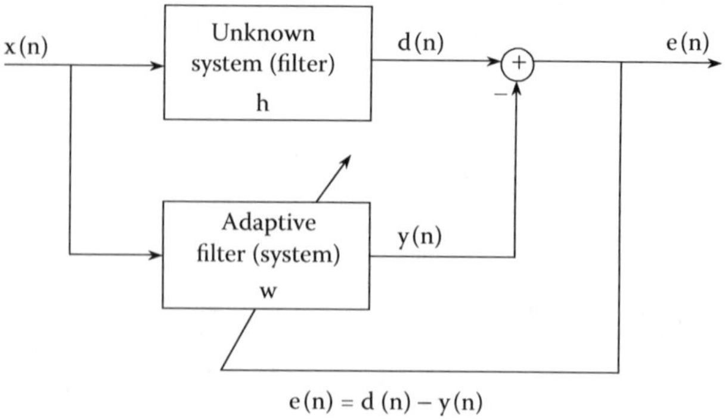

Figure 1.2.1 System identification.

Let us assume that the unknown system is time invariant, which indicates that the coefficients of its impulse response are constants and of finite extent (FIR). Hence, we write

$$d(n) = \sum_{k=0}^{N-1} h_k\, x(n-k) \tag{1.2.1}$$

The output of an adaptive FIR filter with the same number of coefficients, *N*, is given by

$$y(n) = \sum_{k=0}^{N-1} w_k x(n-k) \tag{1.2.2}$$

For these two systems to be equal, the difference $e(n) = d(n) - y(n)$ must be equal to zero. Under these conditions the two sets of coefficients are equal. It is the method of adaptive filtering that will enable us to produce an error, $e(n)$, approximately equal to zero and, therefore, will identify that w_k's $\cong$ h_k's.

1.3 *Outline of the text*

Our purpose in this text is to present the fundamental aspects of adaptive filtering and to give the reader the understanding of how an algorithm, LMS, works for different types of applications. These applications include system identification, noise reduction, echo cancellation during telephone conversation, inverse system modeling, interference canceling, equalization, spectrum estimation, and prediction. In order to aid the reader in his or her understanding of the material presented in this book, an extensive number of MATLAB functions were introduced. These functions are identified with the words "Book MATLAB Function." Ample numbers of examples and figures are added in the text to facilitate the understanding of this particular important signal processing technique. At the end of each chapter, except this introductory chapter, we provided many problems and either complete solutions or hints and suggestions for solving them.

We have tried to provide all the needed background for understanding the idea of adaptive filters and their uses in practical applications. Writing the text, we assumed that the reader will have knowledge at the level of a bachelor's degree in electrical engineering. Although only a small amount of new results is included in this text, its utility of the presented material should be judged by the form of presentation and the successful transferring of the fundamental ideas of adaptive filtering and their use in different areas of research and development.

To accomplish the above mentioned goals, we have started introducing digital signals and their representation in the frequency domain and z-transform domain in Chapter 2. Next, we present in block diagram form the three fundamental discrete systems: finite impulse response (FIR), infinite impulse response (IIR), and the combined system known as the autoregressive mean average (ARMA).

Since most of the input signals in applications of adaptive filtering are random signals, we introduce the notion of random variables, random sequences, and stochastic processes in Chapter 3. Furthermore, we introduce the concepts, and the approaches of finding the power spectral density of random signals.

Chapter 4 develops the foundation for determining minimum mean-square error (MSE) filters. The chapter introduces the Wiener filter, and the "bowl-shaped" error surface. The Wiener filter is also used in a special configuration named self-correcting filtering.

Since the magnitude of the difference between the maximum and minimum value of the eigenvalues of the correlation matrix plays an important role in the rate of convergence of adaptation, Chapter 5 introduces the theory and properties of the eigenvalues and the properties of the error surface.

Chapter 6 introduces the following two gradient search methods: the Newton method and the steepest descent method. A derivation of the convergence properties of the steepest descent method is presented, as well as the valuable geometric analogy of finding the minimum point of the "bowl-shaped" error surface.

Chapter 7 introduces the most celebrated algorithm of adaptive filtering, the LMS algorithm. The LMS algorithm approximates the method of steepest descent. In addition, many examples are presented using the algorithm in diverse applications, such as communications, noise reduction, system identification, etc.

Chapter 8 presents a number of variants of the LMS algorithm, which have been developed since the introduction of the LMS algorithm.

The last chapter, Chapter 9, covers the least squares and recursive least squares signal processing.

Finally, an Appendix was added to present elements of matrix analysis.

chapter 2

Discrete-time signal processing

2.1 Discrete-time signals

Discrete-time signals are seldom found in nature. Therefore, in almost all cases, we will be dealing with the digitization of continuous signals. This process will produce a sequence $\{x(nT)\}$ from a continuous signal $x(t)$ that is sampled at equal time distance T. The *sampling theorem* tells us that, for signals that have a finite spectrum (band-limited signals) and whose highest frequency is ω_N (known as the *Nyquist* frequency), the sampling frequency ω_s must be twice as large as the Nyquist frequency or, equivalently, the sampling time T must be less than one half of its Nyquist time, $2\pi/\omega_N$. In our studies we will consider that all the signals are band-limited. This is a reasonable assumption since we can always pass them through a lowpass filter (pre-filtering). The next section discusses further the frequency spectra of sampled functions.

Basic discrete-time signals

A set of basic continuous and the corresponding discrete signals are included in Table 2.1.1.

Table 2.1.1 Continuous and Discrete-Time Signals

Delta Function

$$\delta(t) = 0 \qquad t \neq 0$$

$$\int_{-\infty}^{\infty} \delta(t)dt = 1$$

$$\delta(nT) = \begin{cases} 1 & n = 0 \\ 0 & n \neq 0 \end{cases}$$

(Continued)

Table 2.1.1 Continuous and Discrete-Time Signals (Continued)

Unit Step Function

$$u(t) = \begin{cases} 1 & t \geq 0 \\ 0 & t < 0 \end{cases}$$

$$u(nT) = \begin{cases} 1 & n \geq 0 \\ 0 & n < 0 \end{cases}$$

The Exponential Function

$$x(t) = e^{-at}u(t)$$

$$x(nT) = e^{-anT}u(nT) = (e^{-aT})^n u(nT) = b^n u(nT)$$

2.2 *Transform-domain representation of discrete-time signals*

Discrete-time Fourier transform (DTFT)

Any non-periodic discrete-time signal $x(t)$ with finite energy has a discrete-time Fourier transform (DTFT), which is found by an approximation of the Fourier transform. The transform is found as follows:

$$X(e^{j\omega T}) = F\{x(nT)\} \cong \int_{-\infty}^{\infty} x(t)e^{-j\omega t}dt = \sum_{n=-\infty}^{\infty} \int_{nT-T}^{nT} x(t)dt$$
$$\cong T\sum_{n=-\infty}^{\infty} x(nT)e^{-j\omega nT} \tag{2.2.1}$$

where the exact integral from nT – T to nT has been replaced by the approximate area T × $x(n\mathrm{T})$. When T = 1, the above equation takes the form

$$X(e^{j\omega}) = \sum_{n=-\infty}^{\infty} x(n)e^{-j\omega n} \tag{2.2.2}$$

The relation $X(e^{j(\omega+2\pi)}) = X(e^{j\omega}e^{j2\pi}) = X(e^{j\omega})$ indicates that the DTFT produces spectra that are periodic with period 2π.

Example 2.2.1: Plot the magnitude and phase spectra for the time function $x(n) = 0.9^n u(n)$.

Solution: The DTFT of x(n) is given by

$$U(e^{j\omega}) = \sum_{n=0}^{\infty} 0.9^n e^{-j\omega n}$$

$$= \sum_{n=0}^{\infty} (0.9e^{-j\omega n}) = \frac{1}{1-0.9e^{-j\omega}}$$

$$= \frac{1}{(1-0.9\cos\omega)+j0.9\sin\omega}$$

$$= \frac{1}{\sqrt{(1-0.9\cos\omega)^2+(0.9\sin\omega)^2}\, e^{j\tan^{-1}(0.9\sin\omega/(1-0.9\cos\omega))}} = A(\omega)e^{j\varphi(\omega)}$$

$$A(\omega) = 1/(\sqrt{(1-0.9\cos\omega)^2+(0.9\sin\omega)^2}),$$

$$\varphi(\omega) = -\tan^{-1}(0.9\sin\omega/(1-0.9\cos\omega))$$

where A(ω) is the magnitude spectrum and $\varphi(\omega)$ is the phase spectrum. In the development we used the Euler's identity $e^{\pm j\theta} = \cos(\theta) \pm j\sin(\theta)$. These two spectra are shown in Figure 2.2.1. The plots are shown only for the period $-\pi < \omega < \pi$, which is the standard presentation of the spectra for digital signals.

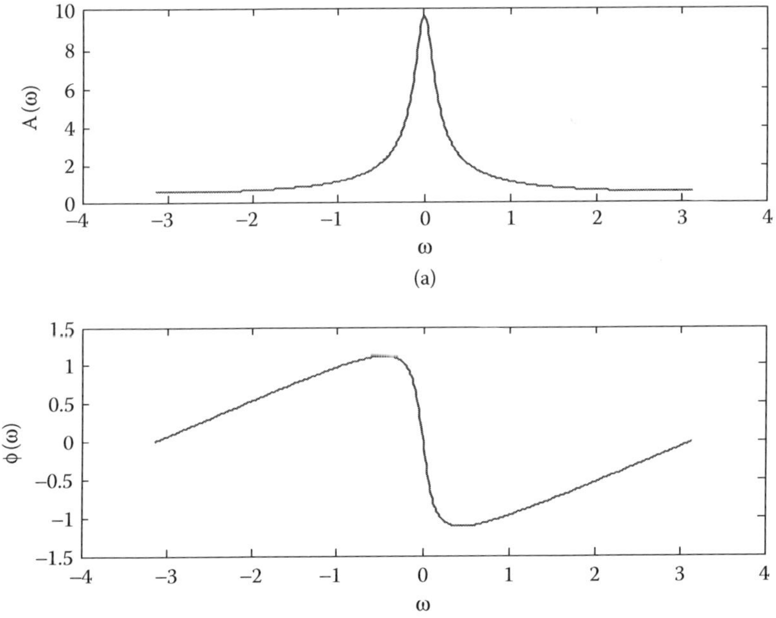

(a)

(b)

Figure 2.2.1 Illustration of Example 2.2.1.

The discrete Fourier transform (DFT)

The N-point discrete Fourier transform (DFT) of a finite signal $x(t)$ sampled at times T apart is given by

$$X\left(k\frac{2\pi}{NT}\right)=T\sum_{n=0}^{N-1}x(nT)e^{-jn\frac{2\pi}{N}k} \quad k=0,1,2,\cdots,N-1 \tag{2.2.3}$$

The inverse DFT is given by

$$x(nT)=\frac{1}{NT}\sum_{k=0}^{N-1}X(k\frac{2\pi}{NT})e^{jk\frac{2\pi}{N}n} \quad n=0,1,2,\cdots,N-1 \tag{2.2.4}$$

For T = 1, the above equations become

$$\begin{aligned} X\left(k\frac{2\pi}{N}\right)&=\sum_{n=0}^{N-1}x(n)e^{-jn\frac{2\pi}{N}k} \quad k=1,2,\cdots,N-1 \\ x(n)&=\sum_{k=0}^{N-1}X\left(k\frac{2\pi}{N}\right)e^{jk\frac{2\pi}{N}n} \quad n=1,2,\cdots,N-1 \end{aligned} \tag{2.2.5}$$

If we replace k in (2.2.5) by $k+N$ we find that X(k) is periodic with period N. Similarly, if we introduce $n + N$ instead of n in the same equation, we find that $x(n)$ is also periodic with period N. This indicates that the DFT is associated with periodic sequences in both time and frequency domain. To obtain the DFT of a sequence and the inverse DFT (IDFT), we use the following MATLAB functions:

```
X=fft(x,N);
x=ifft(X,N);
```

X is an N-point sequence (vector) and x is the input time sequence (vector). The middle point $N/2$ of X is known as the fold-over frequency. Plotting X vs. frequency (vs. $k2\pi/N$), where $2\pi/N$ is the bin frequency, we find that π is the middle point and the frequency spectrum from π to 2π is a reflection of the spectrum from 0 to π. Similarly we find that π/T is the fold over frequency when T is different than one. This indicates that to approximate the spectrum of a continuous signal close enough, we must use sampling time T that is very small.

Example 2.2.2: Find the DFT of the signal $x(t) = \exp(-t)u(t)$.

Solution: First we observe that we cannot take the DFT of the signal from t = 0 to infinity. Because of this limitation, let us find the DFT of the signal from t = 0 to t = 4 and with sampling time T = 1 and T = 0.5. The results will be compared with the exact FT of the signal. The FT of the signal is given by

$$X(\omega)=\int_{-\infty}^{\infty}\exp(-t)u(t)\exp(-j\omega t)dt=\int_{0}^{\infty}\exp(-(1+j\omega)t)dt$$

$$=\frac{1}{-(1+j\omega)}[\exp(-(1+j\omega)t)]_0^{\infty}$$

$$=-\frac{1}{1+j\omega}[0-1]=\frac{1}{1+j\omega}=\frac{1}{(1+\omega^2)^{1/2}}e^{-j\tan^{-1}(\omega)}=A(\omega)e^{j\varphi(\omega)}$$

where the polar form of the complex variable $1+j\omega=\sqrt{1^2+\omega^2}\exp[j\tan^{-1}(\frac{\omega}{1})]$ was used. To find the DFT of the signal for T = 1, we used the following Book MATLAB program:

```
t=0:1:4;
x=exp(-t);
dftx1=fft(x,32); % asked for 32 spectrum bins, MATLAB will
   %pad 32-5 zeros the vector x;
w=0:2*pi/32:2*pi-(2*pi/32);
subplot(211);stem(w,abs(dftx1));
FT1=1./sqrt(1+w.^2);hold on;stem(w,abs(FT1),'filled');
xlabel('\omega');ylabel('Magnitudes of FT and DFT');
title('(a)');axis([0 2*pi 0 2]);legend('DFT','FT');
```

Next, we used the same Book MATLAB program but with a sampling time $T = 1/2$. Hence, the following changes were introduced into the program:

```
nt=0:0.5:4;
x=exp(-nt);
dftx2=0.5*fft(x,32);
w=0:4*pi/32:4*pi-(4*pi/32);
subplot(212);stem(w,abs(dftx2));
FT2=1./sqrt(1+w.^2);
hold on;stem(w,abs(FT2),'filled');
xlabel('\omega');ylabel('Magnitudes of FT and DFT');
title('(b)');axis([0 4*pi 0 1.5]);legend('DFT','FT');
```

Figure 2.2.2a indicates the following: (a) The amplitude of the FT of the continuous signal is constantly decreasing as it should be. (b) The amplitude of the DFT of $x(t)$, for the range $0 \le t \le 4$ has a folding frequency $\pi/1$ (T = 1 in this case). (c) The approximation by the DFT to the

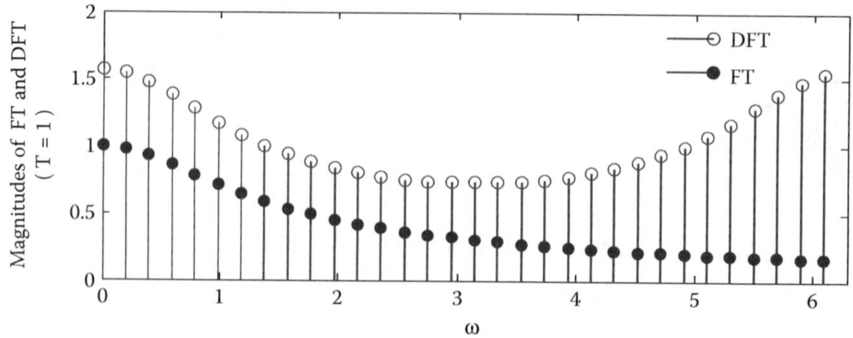

(a)

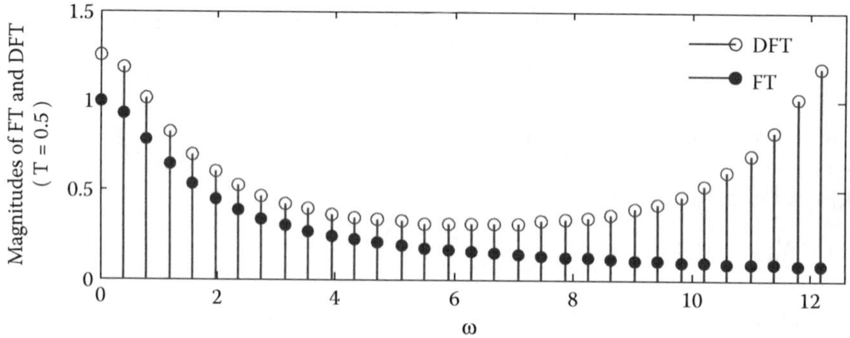

(b)

Figure 2.2.2 Illustration of Example 2.2.2.

exact spectrum is up to the folding frequency. (d) For this set of conditions there is a considerable difference in the two spectra. (e) If the DFT command was fft(x,64), we would have found that the total points for the same frequency range would have been twice as many but the accuracy would have remained the same. (f) If, however, we had increased the time range greater than 4, then the accuracy would have improved.

Figure 2.2.2b shows improvement to the accuracy of the spectrum due to decreasing of the sampling interval to T = 0.5 (increasing the sampling frequency, $\omega_s = 2\pi/T$). If, in addition, we had increased the time range greater than 4, the approximation would have been even better. However, since the largest frequency of this signal is infinite, no finite value of T would have been able to produce the exact spectrum. If, on the other hand, we had band-limited signal with its largest frequency ω_N (Nyquist frequency), then we would have been able to sample the time function at twice its Nyquist frequency, and the two spectra would have been identical.

Note that, although the spectrum of the continuous signal is continuous, we have only plotted the values of the continuous spectrum at the same frequency bins as the discrete one.

2.3 The Z-Transform

The z-transform of a sequence is a transformation of a discrete-time function from its time domain to its frequency domain (and/or z-domain). It is also a powerful tool to study linear time-invariant discrete systems. The z-transform is defined by the following equation (see also Problem 2.3.1):

$$X(z) = Z\{x(n)\} = \sum_{n=-\infty}^{\infty} x(n)z^{-n} \tag{2.3.1}$$

The above equation and its inverse constitute a pair. However, when the inverse of the transformed function is needed, we will use tables.

Example 2.3.1: Find the z-transform of the function $x(n) = 0.9^n u(n)$.

Solution: The z-transform is given by

$$X(z) = \sum_{n=0}^{\infty} 0.9^n z^{-n} = \sum_{n=0}^{\infty} (0.9z^{-1})^n = 1 + 0.9z^{-1} + (0.9z^{-1})^2 + \cdots = \frac{1}{1-0.9z^{-1}} \tag{2.3.2}$$

where we used the geometric series property $(1 + x + x^2 + \ldots) = 1/(1 - x)$ if $|x| < 1$. For the sequence to be summed, the following inequality must be satisfied

$$|0.9z^{-1}| < 1 \quad or \quad |z| > 0.9$$

The absolute value of z is the distance from the origin of the complex plane to a circle on the same plane with radius $|z|$. The whole region $|z| > 0.9$ of the z-plane for this example is known as the region of convergence (ROC). The ROC is important if we are asked to find the inverse z-transform by integration in the complex plane. If we set $z = \exp(j\omega)$ in (2.3.1), we obtain the DTFT for the time function $x(n)$.

There exists a family of z-transforms for which X(z)'s are rational functions of z or, equivalently, z^{-1}. The roots of the numerator and denominator of a rational function are known as the *zeros* and *poles*, respectively.

Table 2.3.1 Properties of the Z-Transform

Property	Time domain	Z-domain
Notation	$x(n)$	$X(z)$
	$x_1(n)$	$X_1(z)$
	$x_2(n)$	$X_2(n)$
Linearity	$ax_1(n)+bx_2(n)$	$aX_1(z)+bX_2(z)$
Time shifting	$x(n-k)$	$z^{-k}X(z)$
Scaling	$a^n x(n)$	$X(a^{-1}z)$
Conjugation	$x^*(n)$	$X^*(z^*)$
Differentiation	$nx(n)$	$-z(dX(z)/dz)$
Convolution	$x(n)*h(n)$	$X(z)H(z)$
Initial value (if a sequence is zero for $n < n_0$)		$x(n_0) = z^{n_0} X(z)\big\|_{z\to\infty}$
Final value (if $x(\infty)$ exists)		$\lim_{n\to\infty} x(n) = \lim_{z\to 1}(1-z^{-1})X(z)$

For example, the rational function in the above example has a zero at 0 and a pole at 0.9. The location of the poles and zeros in the complex plane play an important role in the study of discrete signals and systems. Some fundamental properties and some pairs of *z*-transforms are given in Table 2.3.1 and Table 2.3.2, respectively.

Table 2.3.2 Z-Transform Pairs

Time domain	Z-domain	ROC
$\delta(n)$	1	all z
$\delta(n-k)$	z^{-k}	all z
$u(n)$	$\dfrac{z}{z-1}$	$\|z\| > 1$
$nu(n)$	$\dfrac{z}{(z-1)^2}$	$\|z\| > 1$
$a^n u(n)$	$\dfrac{z}{z-a}$	$\|z\| > a$
$cos(n\omega T)u(n)$	$\dfrac{z^2 - z\cos\omega T}{z^2 - 2z\cos\omega T + 1}$	$\|z\| > 1$
$sin(n\omega T)u(n)$	$\dfrac{z\sin\omega T}{z^2 - 2z\cos\omega T + 1}$	$\|z\| > 1$
$na^n u(n)$	$\dfrac{z}{(z-a)^2}$	$\|z\| > a$

2.4 Discrete-time systems

A discrete time-invariant system is a physical device or an algorithm that transforms an input (or excitation) signal $\{x(n)\}$ into another one, called the output signal, $\{y(n)\}$. Every discrete system is defined by its response, known as the impulse response $h(n)$ of the system, to a unit impulse input $\delta(n)$, which is defined as follows:

$$\delta(n) = \begin{cases} 1 & n = 0 \\ 0 & n \neq 0 \end{cases} \tag{2.4.1}$$

The relationship that provides the output of a system, given its input, is a mathematical relationship that maps the input to the output. This relationship is given by the convolution operation

$$y(n) = x(n) * h(n) = \sum_{m=-\infty}^{\infty} x(m)h(n-m) = \sum_{m=-\infty}^{\infty} h(m)x(n-m) \tag{2.4.2}$$

The above equation indicates the following operations required to find the output $y(n)$. (a) We select another domain m for $x(n)$ and $h(n)$. (b) In this domain, we represent one of the functions as it was in the n domain by simply substituting each n with m. (c) We flip the other function in m domain (see the minus sign in front of m) and shift it by n. (d) Next, we first multiply the two sequences term by term, as they were arranged by the shifting process, and then add the results. (e) The result is the output of the system at time n. (f) We repeat the same procedure for all n's and, thus, we find the output for all times from minus infinity to infinity.

Example 2.4.1: Find the convolution of the following two functions: $f(n) = u(n)$ and $h(n) = a^n u(n)$, $|a| < 1$.

Solution: If we flip and shift the unit step function, we observe that when $n < 0$, the two functions do not overlap and, hence, the output is zero. Therefore, we find the output only if n is equal to or greater than zero. Hence, (2.4.2) for this case takes the form

$$y(n) = \sum_{m=0}^{\infty} u(n-m)a^m = \sum_{m=0}^{n} a^m = 1 + a + a^2 + \cdots + a^n = \frac{1-a^{n+1}}{1-a}$$

where we used the formula for a finite geometric series. The step function $u(n-m)$ is zero for $m > n$ and is equal to one for $m \leq n$.

The two most important properties of the convolution operation are:

Linearity:

$$g(n) = (f(n) + h(n)) * y(n) = f(n) * y(n) + h(n) * y(n) \tag{2.4.3}$$

z-transform:

$$Z\{f(n) * h(n)\} = Z\{h(n) * f(n)\} = F(z)H(z) \tag{2.4.4}$$

Transform-domain representation

Based on the above development, the output $y(n)$ in the z-domain of any system having impulse response $h(n)$ and an input $x(n)$ is given by

$$Y(z) = H(z)X(z) \tag{2.4.5}$$

H(z) is known as the system function and plays a fundamental role in the analysis and characterization of LTI systems. If the poles of H(z) are inside the unit circle, the system is stable and $H(e^{j\omega})$ provides its frequency response.

A realizable discrete LTI and causal (its impulse response is zero for $n < 0$) system can be represented by the following linear difference equation

$$y(n) + \sum_{m=0}^{p} a(m)y(n-m) = \sum_{m=0}^{q} b(m)x(n-m) \tag{2.4.6}$$

To find the output of a discrete system, we use the MATLAB function

```
y=filter(b,a,x);%b=row vector of the b's; a=row vector of the a's;
%x=row vector of the input {x(n)}; y=output vector;
```

Taking the z-transform of (2.4.6), and remembering that the transform is with respect to n and that the functions $h(.)$ and $x(.)$ are shifted by n, we obtain the following relation

$$H(z) = \sum_{m=0}^{\infty} h(m)z^{-m} = \frac{Y(z)}{X(z)} = \frac{\sum_{m=0}^{q} b(m)z^{-m}}{1 + \sum_{m=1}^{p} a(m)z^{-m}} = \frac{B(z)}{A(z)} \tag{2.4.7}$$

The above system is known as the Autoregressive Moving Average (ARMA) system. If a's are zero, then the system in time and in z-domain is given, respectively, by

$$y(n) = \sum_{m=0}^{q} b(m)x(n-m)$$
$$H(z) = B(z) \tag{2.4.8}$$

The above system is known as the Finite Impulse Response (FIR) or nonrecursive. If the b's are zero besides the $b(0)$, then the system in time and z-domain is given by

$$y(n) + \sum_{m=0}^{p} a(m)x(n-m) = b(0)x(n)$$
$$H(z) = \frac{b(0)}{1 + \sum_{m=0}^{p} a(m)z^{-m}} = \frac{b(0)}{A(z)} \tag{2.4.9}$$

The above system is known as the Infinite Impulse Response (IIR) or recursive.

Figure 2.4.1 shows the above three systems in their filter realization and of the second order each. In this text we will be mostly concerned with the FIR filters, since they are inherently stable.

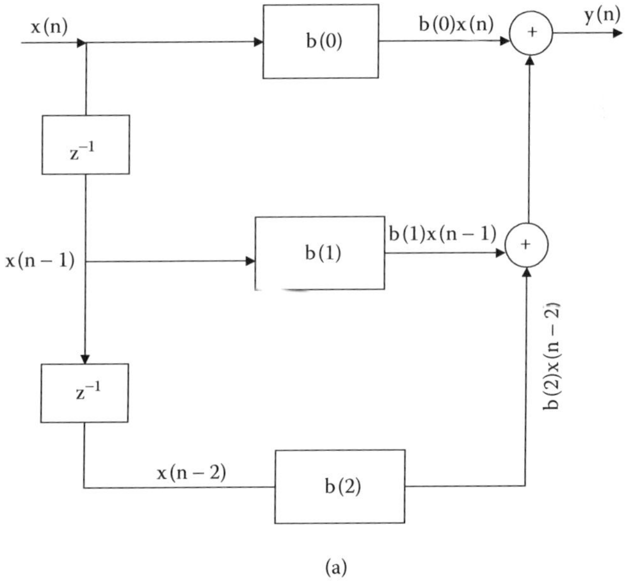

Figure 2.4.1(a) FIR system.

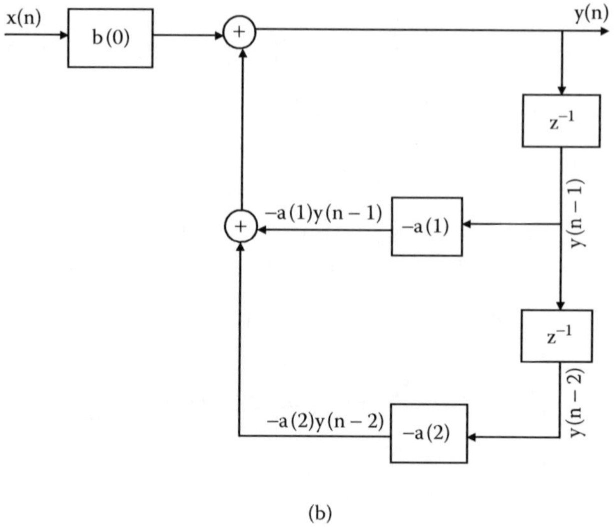

Figure 2.4.1(b) IIR system.

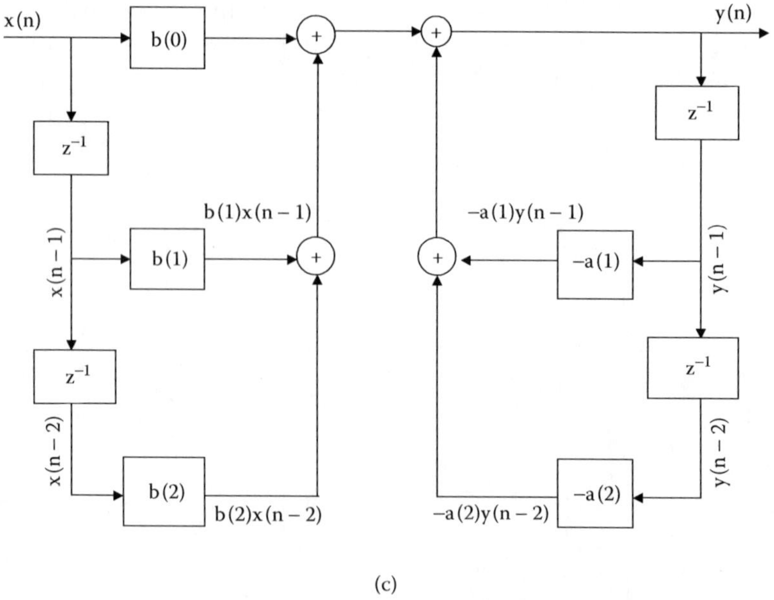

Figure 2.4.1(c) ARMA system.

Problems

2.2.1 Compare the amplitude spectrum at $\omega = 1.6$ rad/s between the continuous function $x(t) = \exp(-|t|)$ and DTFT of the same signal with sampling interval T = 1s and T = 0.1s.

2.2.2 Repeat Problem 2.2.1 using the DFT. Assume the range for the DFT is $-6 \le t \le 6$, and use sampling intervals T = 1 and T = 0.2 and $\omega = \pi/2$.

2.3.1 Find the z-transform of the functions (see Table 2.3.2): (a) $\cos(n\omega T)u(n)$; (b) $na^n u(n)$.

2.4.1 The transfer function of a discrete system is $H(e^{j\omega T}) = 1/(1 - e^{-j\omega T})$. Find its periodicity and sketch its amplitude frequency response for T = 1 and T = 1/2. Explain your results.

2.4.2 Find the type (lowpass, highpass, etc.) of the filter $H(z) = \dfrac{a + bz^{-1} + cz^{-2}}{c + bz^{-1} + az^{-2}}$.

Hints-solutions-suggestions

2.2.1:

$$X(\omega) = \int_{-\infty}^{\infty} e^{-|t|} e^{-j\omega t} dt = \int_{-\infty}^{0} e^{-(j\omega - 1)t} dt + \int_{0}^{\infty} e^{-(j\omega+1)t} dt$$

$$= \frac{1}{-(j\omega - 1)} [e^{-(j\omega-1)t} \Big|_{-\infty}^{0}] + \frac{1}{-(j\omega + 1)} [e^{-(j\omega+1)t} \Big|_{0}^{\infty}]$$

$$= \frac{1}{-(j\omega - 1)} + \frac{1}{(j\omega + 1)} = \frac{2}{(1 + \omega^2)} \Rightarrow X(1.6) = \frac{2}{1 + 1.6^2} = 0.5618.$$

$$x(nT) = e^{-|nT|} \Rightarrow X(e^{j\omega T}) = T \sum_{n=-\infty}^{\infty} e^{-|nT|} e^{-j\omega nT} = T \sum_{n=-\infty}^{-1} e^{nT} e^{-j\omega nT}$$

$$+ T \sum_{n=0}^{\infty} e^{-nT} e^{-j\omega nT} = -T + T \sum_{n=0}^{\infty} (e^{-T + j\omega T})^n + T \sum_{n=0}^{\infty} (e^{-T - j\omega T})^n = -T$$

$$+ T + \frac{T}{1 - e^{-T} e^{j\omega T}} + \frac{T}{1 - e^{-T} e^{-j\omega T}} = -T + T \frac{1 - e^{-T} e^{-j\omega T} + 1 - e^{-T} e^{j\omega T}}{1 - e^{-T} e^{j\omega T} - e^{-T} e^{-j\omega T} + e^{-2T}}$$

$$= -T + T \frac{2 - 2e^{-T} \cos \omega T}{1 - 2e^{-T} \cos \omega T + e^{-2T}} \Rightarrow X(e^{j1.6 \times 1}) = 0.7475, X(e^{j1.6 \times 0.1}) = 0.5635$$

2.2.2:

Use the help of MATLAB for the frequency $\pi/2$. (a) T = 1; n = –6:1:6; x = exp(–abs(n)); X = 1*fft(x, 500). The frequency bin is $2\pi/(500-1)$ and hence $k(2\pi/500)=\pi/2$, which implies that $k = 125$ and, hence, abs(X(125)) = 0.7674. (b) Similarly, for T = 0.2 we find $k = 25$ and abs(X(125)) = 0.6194. The exact value is equal to $2/(1+(\pi/2)^2) = 0.5768$.

2.3.1:

$$a)\quad \sum_{n=0}^{\infty}\cos n\omega T\, z^{-n}=\sum_{n=0}^{\infty}\frac{e^{jn\omega T}+e^{jn\omega T}}{2}z^{-n}=\frac{1}{2}\sum_{n=0}^{\infty}(e^{j\omega T}z^{-1})^n+\frac{1}{2}\sum_{n=0}^{\infty}(e^{-j\omega T}z^{-1})^n$$

$$=\frac{1}{2}\frac{1}{1-e^{j\omega T}z^{-1}}+\frac{1}{2}\frac{1}{-e^{-j\omega T}z^{-1}}=\frac{z^2-z\cos\omega T}{z^2-2z\cos\omega T+1}$$

$$b)\quad \sum_{n=0}^{\infty}na^nz^{-n}=\sum_{n=0}^{\infty}a^n\frac{dz^{-n}}{dz}(-nz)=-z\frac{d}{dz}\sum_{n=0}^{\infty}a^nz^{-n}=-z\frac{d}{dz}\left(\frac{z}{z-a}\right)=\frac{z}{(z-a)^2}$$

2.4.1:

By setting $\omega=\omega+(2\pi/T)$ we find that $H(e^{j\omega T})$ is periodic with period $2\pi/2$. From the plots of $|H(e^{j\omega T})|^2$ we observe that the fold-over frequency is at π/T.

2.4.2:

All-pass

chapter 3

Random variables, sequences, and stochastic processes

3.1 Random signals and distributions

Most signals in practice are not deterministic and can not be described by precise mathematical analysis, and therefore we must characterize them in probabilistic terms using the tools of statistical analysis.

A *discrete random signal* $\{X(n)\}$ is a sequence of indexed random variables (rv's) assuming the values:

$$\{x(0), x(1), x(2), ...\} \tag{3.1.1}$$

The random sequence with values $\{x(n)\}$ is discrete with respect to sampling index n. Here we will assume that the random variable at any time n is a continuous function, and therefore, it is a continuous rv at any time n. This type of sequence is also known as *time series*.

A particular rv, $X(n)$, is characterized by its *probability density function* (pdf) $f(x(n))$

$$f(x(n)) = \frac{\partial F(x(n))}{\partial x(n)} \tag{3.1.2}$$

and its *cumulative density function* (cdf) $F(x(n))$

$$F(x(n)) = p(X(n) \le x(n)) = \int_{-\infty}^{x(n)} f(y(n))dy(n) \tag{3.1.3}$$

$p(X(n) \le x(n))$ is the probability that the rv $X(n)$ will take values less than or equal to $x(n)$ at time n. As the value of $x(n)$ goes to infinity, $F(x(n))$ approaches unity. Similarly, the *multivariate* distributions of rv's are given by

$$F(x(n_1),\cdots,x(n_k)) = p(X(n_1) \le x(n_1),\cdots,X(n_k) \le x(n_k))$$

$$f(x(n_1),\cdots,x(n_k)) = \frac{\partial^k F(x(n_1),\cdots,x(n_k))}{\partial x(n_1)\cdots \partial x(n_k)} \tag{3.1.4}$$

Note that here we have used a capital letter to indicate rv. In general, we shall not keep this notation, since it will be obvious from the context.

To obtain a formal definition of a discrete-time stochastic process, we consider an experiment with a finite or infinite number of unpredictable outcomes from a sample space, $S(z_1, z_2, \ldots)$, each one occurring with a probability $p(z_i)$. Next, by some rule we assign a deterministic sequence $x(n, z_i)$, $-\infty < n < \infty$, to each element z_i of the sample space. The sample space, the probabilities of each outcome, and the sequences constitute a *discrete-time stochastic process* or *random sequence*. From this definition we obtain the following four interpretations:

- $x(n,z)$ is an rv if n is fixed and z is variable.
- $x(n,z)$ is a sample sequence called realization if z is fixed and n is variable.
- $x(n,z)$ is a number if both n and z are fixed.
- $x(n,z)$ is a stochastic process if both n and z are variables.

Each time we run an experiment under identical conditions, we create a sequence of rv's $\{X(n)\}$, which is known as a realization and constitutes an event. A realization is one member of a set called the ensemble of all possible results from the repetition of an experiment.

Book MATLAB script file

To create a script file, we first create the file, as shown below, as a new MATLAB file. Then we save the file named 'realizations.m', for example, in the directory c:\aamatlab. When we are in the MATLAB command window, we attach the above directory to MATLAB path using the following command: path(path,'c:\aamatlab'). Then the only thing we have to do is to write: realizations and automatically the MATLAB will produce Figure 3.1.1, which shows four realizations of a stochastic process with zero mean value.

```
%script file: realizations
for n=1:4
x(n,:)=rand(1,50)-0.5;%produces matrix x with 4 rows
%and 50 columns of zero mean white noise;
end;
m=0:49;
for i=1:4
subplot(4,1,i);stem(m,x(i,:),'k');%plots four rows of matrix x;
end;
```

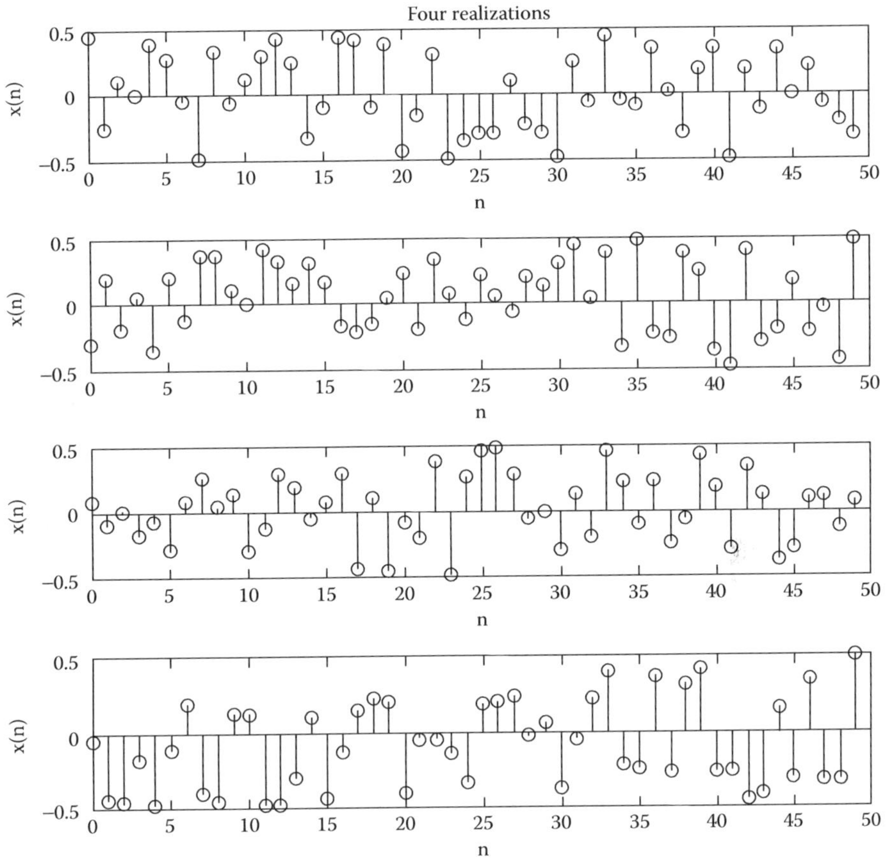

Figure 3.1.1 Four realizations of zero mean white noise.

Stationary and ergodic processes

It is seldom in practice that we will be able to create an ensemble of a random process with numerous realizations so that we can find some of its statistical characteristics, e.g., mean value, variance, etc. To find these statistical quantities we need the pdf of the process, which, most of the time, is not possible to produce. Therefore, we will restrict our studies to processes that are easy to study and easy to handle mathematically.

The process that produces an ensemble of realizations and whose statistical characteristics do not change with time is called *stationary*. For example, the pdfs of the rv's $x(n)$ and $x(n + k)$ of the process $\{x(n)\}$ are the same independently of the values of n and k.

Since we are unable to produce ensemble averages in practice, we are left with only one realization of the stochastic process. To overcome this difficulty we assume that the process is *ergodic*. This characterization permits us to find the desired statistical characteristics of the process from only one realization at hand. We refer to those statistical values as *sample mean*, *sample variance*, etc.

3.2 Averages

Mean value

The *mean* value or *expectation* value μ_n at time n of a random variable $x(n)$ having a pdf $f(x(n))$ is given by

$$\mu_n = E\{x(n)\} = \int_{-\infty}^{\infty} x(n) f(x(n)) dx(n) \tag{3.2.1}$$

where $E\{\cdot\}$ stands for *expectation* operator. We can also use the ensemble of realizations to obtain the mean value using the frequency interpretation formula

$$\mu_n = \lim_{N\to\infty} \left\{ \frac{1}{N} \sum_{i=1}^{N} x_i(n) \right\} \qquad N = \textit{number of realizations} \tag{3.2.2}$$

where $x_i(n)$ is the ith outcome at sample index n (or time n) of the ith realization. Depending on the type of the rv, the mean value may or may not vary with time.

For an ergodic process, we find the sample mean (estimator of the mean) using the *time-average* formula

$$\hat{\mu} = \frac{1}{N} \sum_{n=0}^{N-1} x(n) \tag{3.2.3}$$

It turns out (see Problem 3.1.2) that the sample mean $\hat{\mu}$ is equal to the population mean and, therefore, we call the sample mean an *unbiased* estimator.

Correlation

The *cross-correlation* between two random sequences is defined by

$$r_{xy}(m,n) = E\{x(m), y(n)\} = \iint x(m)y(n) f(x(m), y(n)) dx(m) dy(n) \tag{3.2.4}$$

where the integrals are from minus infinity to infinity. If $x(n) = y(n)$, the correlation is known as the *autocorrelation*. Having an ensemble of realizations, the frequency interpretation of the autocorrelation function is found using the formula

$$r_x(m,n) = \lim_{N\to\infty} \left\{ \frac{1}{N} \sum_{i=1}^{N} x_i(m) x_i(n) \right\} \tag{3.2.5}$$

Note that we use one subscript for autocorrelation functions. In case we have cross-correlation we will use both subscripts.

Example 3.2.1: Using Figure 3.1.1, find the mean for n = 15 and the autocorrelation function for time difference of five, n = 20 and n = 25.

Solution: The desired values are

$$\mu_{15} = \frac{1}{4}\sum_{i=1}^{4} x_i(15) = \frac{1}{4}(-0.45+0.1+0.2-0.1) = 0.05$$

$$r_x(20,25) = \frac{1}{4}\sum_{i=1}^{4} x_i(20)\,x_i(25) = \frac{1}{4}[(-0.4)(0.2)+(-0.1)\,(0.45)+(0.25)(0.2)$$

$$+(-0.45)(-0.3)] = 0.0150$$

Because the number of realizations is very small, both values found above are not expected to be accurate.

Figure 3.2.1 shows the mean value at 50 individual times and the autocorrelation function for 50 differences (from 0 to 49) known as *lags*.

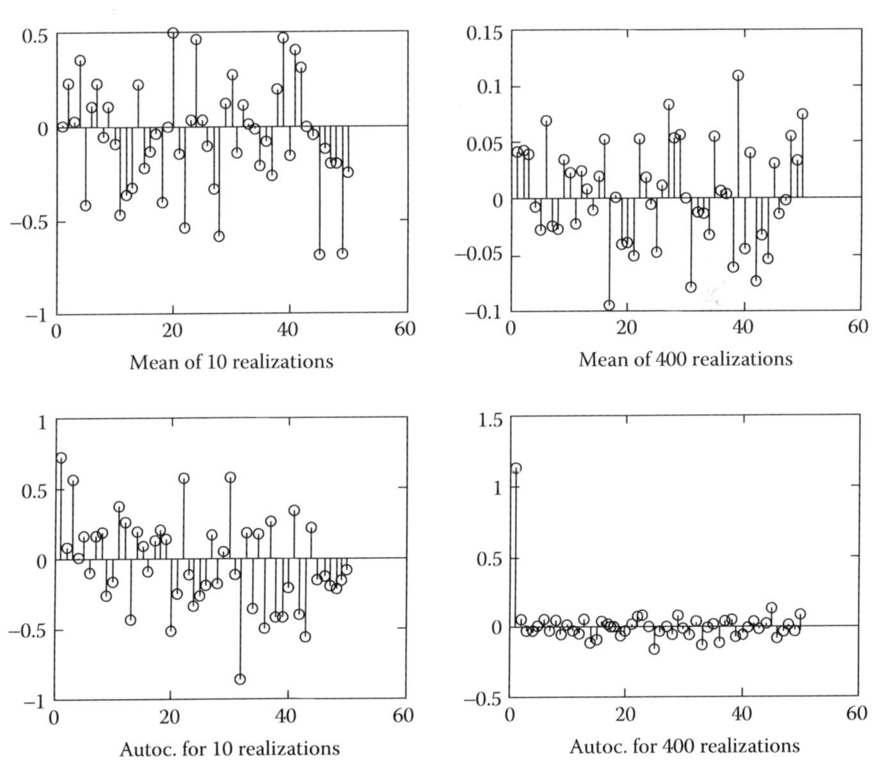

Figure 3.2.1 Means and autocorrelations based on frequency interpretation.

These results were found using the MATLAB function given below. Note that as the number of realizations increases, the mean tends to zero and the autocorrelation tends to a delta function, as it should be, since the random variables are independent (white noise).

Book MATLAB function to find the mean and autocorrelation function using the frequency interpretation formula:

```
function[mx,rx]=aameanautocensemble(M,N)
%function[mx,rx]=aameanautocensemble(M,N);
%we create an MxN matrix;M=number of realizations;
%N=number of time slots;easily modified for other types
%of pdf's;
x=randn(M,N);%randn=gives Gaussian distributed white noise
mx=sum(x,1)/M;%with zero mean;sum(x,1)=sums all the rows;
for i=1:N %sum(x,2)=sums all the columns;
rx(i)=sum(x(:,1).*x(:,i))/M;
end;
```

To find the sample autocorrelation function from one realization we can use the formula

$$\hat{r}(m)=\frac{1}{N-|m|}\sum_{n=0}^{N-|m|-1}x(n)x(n+|m|) \quad m=0,1,\cdots,N-1 \tag{3.2.6}$$

The absolute value of m ensures the symmetry of the sample autocorrelation function at $n = 0$. Although this formula gives an unbiased autocorrelation function, it sometimes produces autocorrelation matrices (discussed below), which do not have inverses. Therefore, it is customary in practice to use any one of the biased formulas

$$\hat{r}(m)=\frac{1}{N}\sum_{n=0}^{N-|m|-1}x(n)x(n+|m|) \quad 0\le m\le N-1$$

or (3.2.7)

$$\hat{r}(m)=\frac{1}{N}\sum_{n=m}^{N-1}x(n)x(n-m) \quad 0\le m\le N-1$$

Book MATLAB function to find the unbiased autocorrelation function:

```
function[r]=aasampleunbisedautoc(x,lg)
%this function finds the unbiased autocorrelation function
%from 0 to lag lg;it is recommended that lg is about 20-30% of N;
```

```
N=length(x);%x=data;
for m=1:lg
for n=1:N+1-m
xs(m,n)=x(n-1+m);
end;
end;
r1=xs*x';
for m=1:lg
den(m)=N+1-m;
end;
r=r1'./den;
```

Book MATLAB function to find the biased autocorrelation function:

```
function[r]=aasamplebiasedautoc(x,lg)
%this function finds the biased autocorrelation function
%with lag from 0 to lg; it is recommended that lg is 20-30% of
%N;
N=length(x);%x=data;lg=lag;
for m=1:lg
for n=1:N+1-m
xs(m,n)=x(n-1+m);
end;
end;
r1=xs*x';
r=r1'./N
```

We can also use MATLAB function to obtain the biased or unbiased sample autocorrelation and cross-correlation. The function is:

```
r=xcorr(x,y,'biased');     % for the biased cased, and
r=xcorr(x,y,'unbiased');   % for the unbiased case.
% x,y are N length vectors; r is a 2N-1 symmetric
% cross-correlation vector;
% in case the vectors do not have the same length, the
% shorter one will be zero-padded;
```

Note: If none of the options (i.e., *biased* or *unbiased*) is used, the default value is *biased* and the result will not be divided by N.

The reader is encouraged to find several interesting options in using the **xcorr** command by writing **help xcorr** or **doc xcorr** on the MATLAB command window.

Covariance

The *covariance* of a random sequence is defined by

$$
\begin{aligned}
c_x(m,n) &= E\{(x(m)-\mu_m)(x(n)-\mu_n)\} \\
&= E\{x(m)x(n)\} - \mu_m\mu_n = r_x(m,n) - \mu_m\mu_n
\end{aligned}
\tag{3.2.8}
$$

The variance is found by setting $m = n$ in (3.2.8). Hence,

$$c_x(n,n) = \sigma_n^2 = E\left\{(x(n) - \mu_n)^2\right\} = E\left\{x^2(n)\right\} - \mu_n^2$$

If the mean value is zero, then the variance and the correlation function are identical.

$$c_x(n,n) = \sigma_n^2 = E\left\{x^2(n)\right\} = r_x(n,n) \tag{3.2.9}$$

Independent and uncorrelated rv's

If the joint pdf of two rv's can be separated into two pdf's, $f_x(m)f_y(n)$ $f_{x,y}(m,n) = f_x(m)f_y(n)$, then the rv's are statistically *independent*. Hence,

$$E\left\{x(m)x(n)\right\} = E\left\{x(m)\right\}E\left\{x(n)\right\} = \mu_m\mu_n \tag{3.2.10}$$

The above equation is a necessary and sufficient condition for the two random variables $x(m)$, $x(n)$ to be uncorrelated. Note that independent random variables are always uncorrelated. However, the converse is not necessarily true.

If the mean value of any two uncorrelated rv's is zero, then the random variables are called *orthogonal*. In general, two rv's are called orthogonal if their correlation is zero.

3.3 *Stationary processes*

For a *wide-sense* (or weakly) stationary process, the cdf satisfies the relationship

$$F(x(m), x(n)) = F(x(m+k), x(n+k)) \tag{3.3.1}$$

for any m, n, and k. If the above relationship is true for any number of rv's of the time series, then the process is known as *strictly* stationary process.

The basic properties of a wide-sense stationary process are (see Problem 3.3.1):

$$\mu_n(n) = \mu = \text{constant} \qquad r_x(k) = r_x(-k)$$

$$r_x(m,n) = r_x(m-n) \qquad r_x(0) \geq r_x(k) \tag{3.3.2}$$

Autocorrelation matrix

If $\mathbf{x} = [x(0)\ x(1)\ \ldots.\ x(p)]^T$ is a vector representing a finite random sequence then the autocorrelation matrix is given by

$$\mathbf{R}_x = E\{\mathbf{x}\,\mathbf{x}^T\} = \begin{bmatrix} E\{x(0)x(0)\} & E\{x(0)x(1)\} & \cdots & E\{x(0)x(p)\} \\ E\{x(1)x(0)\} & E\{x(1)x(1)\} & \cdots & E\{x(1)x(p)\} \\ \vdots & \vdots & & \vdots \\ E\{x(p)x(0)\} & E\{x(p)x(1)\} & \cdots & E\{x(p)x(p)\} \end{bmatrix}$$

$$= \begin{bmatrix} r_x(0) & r_x(-1) & \cdots & r_x(-p) \\ r_x(1) & r_x(0) & & r_x(-p+1) \\ \vdots & \vdots & & \vdots \\ r_x(p) & r_x(p-1) & \cdots & r_x(0) \end{bmatrix} \tag{3.3.3}$$

Example 3.3.1: Find: (a) the unbiased autocorrelation with lag 20 of a 40 term sequence; (b) the biased autocorrelation with the same settings; and (c) a 4 × 4 autocorrelation matrix. Use a sequence of rv's having Gaussian distribution and zero mean value.

Solution: We use the MATLAB function x = *randn*(1,40) to create the sequence $\{x(n)\}$ of 40 terms, which are white rv's and Gaussian distributed. Then we use the two MATLAB functions, which were given above, ru=aasampleunbiasedautoc(x,20) and rb=aasamplebiasedautoc(x,20) to produce the results shown in Figure 3.3.1. To create the 4 × 4 autocorrelation matrix we use the following MATLAB function: R_x= *toeplitz* (rb(1,1:4)′, rb(1,1:4));.

$$\mathbf{R}_x = \begin{bmatrix} 0.8613 & -0.0363 & -0.0232 & 0.2112 \\ -0.0363 & 0.8613 & -0.0363 & -0.0232 \\ -0.0232 & -0.0363 & 0.8613 & -0.0363 \\ 0.2112 & -0.0232 & -0.0363 & 0.8613 \end{bmatrix} \tag{3.3.4}$$

Note the symmetry of the matrix along the diagonals. This type of matrices is known as *Toeplitz*.

Note: The output vector rb is a row vector, and the parenthesis (1,1:4) instructs MATLAB to take the fist row of rb and the first four columns of that row.

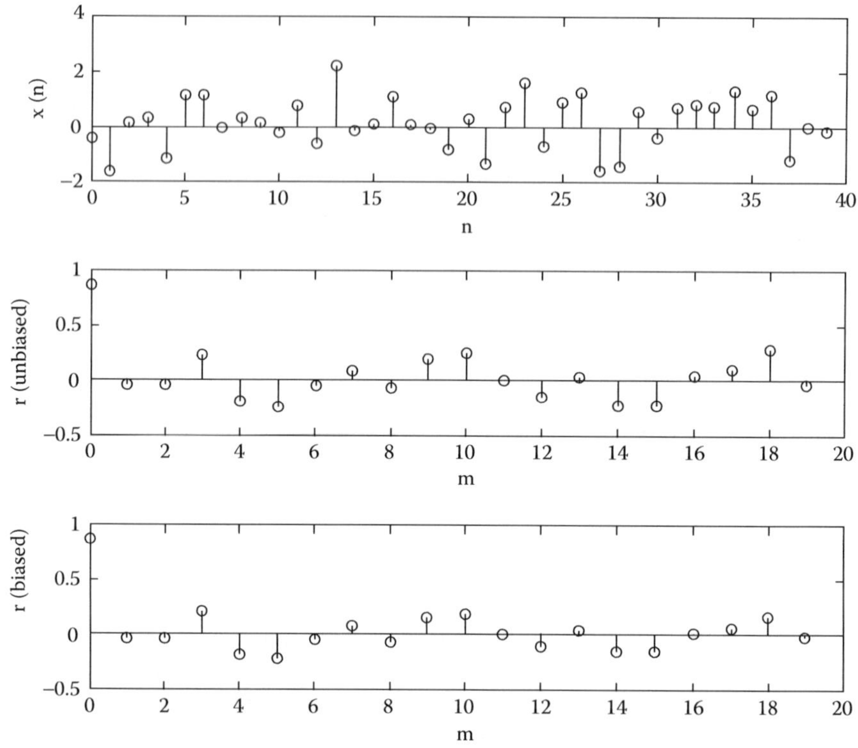

Figure 3.3.1 Illustration of Example 3.3.1.

Note: If we have a row vector **x** and need to create a row vector **y** with elements of **x** from k to m only, we write

```
y=x(1,k:m);
```

If **x** is a column vector, and we need to find a column vector **y** with elements of x from k to m only, we write

```
y=x(k:m,1);
```

Example 3.3.2: Let $\{v(n)\}$ be a zero mean, uncorrelated Gaussian random sequence with variance $\sigma_v^2(n) = \sigma^2$ = constant.

a. Characterize the random sequence $\{v(n)\}$.
b. Determine the mean and the autocorrelation of the sequence $\{x(n)\}$ if $x(n) = v(n) + av(n-1)$, in the range $\infty < n < \infty$ where a is a constant.

Solution: (a) the variance of $\{v(n)\}$ is constant and, hence, is independent of the time, n. Since $\{v(n)\}$ is an uncorrelated sequence, it is also independent due to the fact that it is Gaussian sequence. From (3.2.8) we obtain $c_v(m,n) = r_v(m,n) - \mu_m\mu_n = r_v(m,n)$ or $\sigma^2 = r_v(n,n)$ = constant. Hence, $r_v(m,n) = \sigma^2\delta(m-n)$, which implies that $\{v(n)\}$ is a WSS process. (b) $E\{x(n)\} = 0$ since $E\{v(n)\} = E\{v(n-1)\} = 0$. Hence,

$$r_x(m,n) = E\{[v(m) + av(m-1)][v(n) + av(n-1)]\} = E\{v(m)v(n)\}$$

$$+aE\{v(m-1)v(n)\} + aE\{v(m)v(n-1)\} + a^2E\{v(m-1)v(n-1)\} = r_v(m,n)$$

$$+ar_v(m-1,n) + ar_v(m,n-1) + a^2r_v(m-1,n-1) = \sigma^2\delta(m-n) + a\sigma^2\delta(m-n+1)$$

$$+a^2\sigma^2\delta(m-n) = (1+a^2)\sigma^2\delta(l) + a\sigma^2\delta(l-1) + a\sigma^2\delta(l+1)$$

Since the mean of the process $\{x(n)\}$ is zero (constant), and its autocorrelation is a function of the lag factor $l = m - n$, it is a WSS process.

3.4 Special random signals and probability density functions

White Noise

A WSS discrete random sequence that satisfies the relation

$$f(x(0), x(1), \ldots) = f(x(0))f(x(1))\ldots \tag{3.4.1}$$

is a pure random sequence whose elements $x(n)$ are statistically independent and identically distributed (iid). Therefore, the zero mean iid sequence has the following correlation function

$$r_x(m-n) = E\{x(m)x(n)\} = \sigma_x^2\delta(m-n) \tag{3.4.2}$$

where σ_x^2 is the variance of the signal and $\delta(m-n)$ is the discrete-time delta function. For $m \neq n$, $\delta(m-n) = 0$ and, hence, (3.4.2) becomes

$$r_x(k) = \begin{cases} \sigma_x^2\delta(k) & k = 0 \\ 0 & k \neq 0 \end{cases} \tag{3.4.3}$$

For example, a random process consisting of a sequence of uncorrelated Gaussian rv's is a white noise process referred to as *white Gaussian noise* (WGN).

Gaussian processes

The pdf of a Gaussian rv $x(n)$ at time n is given by

$$f(x(n)) = \frac{1}{\sqrt{2\pi\sigma_n^2}}\exp(-\frac{(x(n)-\mu_n)^2}{2\sigma_n^2}) = N(\mu_n,\sigma_n^2) \tag{3.4.4}$$

Example 3.4.1: Find the joint pdf of a sequence of WGN with n elements; each one having zero mean value and the same variance.

Solution: The joint pdf is

$$\begin{aligned} f(x(1),x(2),\cdots,x(n)) &= f_1(x(1))f_2(x(2))\cdots f_n(x(n)) \\ &= \frac{1}{(2\pi)^{n/2}\sigma_x^n}\exp\left[-\frac{1}{2\sigma_x^2}\sum_{k=1}^{n}x^2(k)\right] \end{aligned} \tag{3.4.5}$$

A discrete-time random process $\{x(n)\}$ is said to be Gaussian if every finite collection of samples of $x(n)$ are jointly Gaussian. A Gaussian random process has the following properties. (a) It is completely defined by its mean vector and covariance matrix. (b) Any linear operation on the time variables produces another Gaussian random process. (c) All higher moments can be expressed by the first and second moments of the distribution (mean, covariance). (d) White noise is necessarily generated by iid samples (independence implies uncorrelated rv's and vice versa).

To produce a WGN with zero mean and unit variance, the following MATLAB function can be used:

```
x=randn(1,N);% x is a row vector with N elements of WGN type
             % with zero mean and unit variance;
```

In case it is desired to change the mean and the variance, we use the following transformation of the vector **x**.

```
z=a*x+m; % the variance of z equals a², and its mean equals m;
```

Exponential distribution

$$f(x) = \begin{cases} \frac{1}{b}\exp(-x/b) & 0 \le x < \infty,\ b > 0 \\ 0 & \text{otherwise} \end{cases} \tag{3.4.6}$$

Algorithm

1. Generate u from a uniform distribution (0,1)
2. $x = -b\ln(bu)$
3. keep x

Book MATLAB function

```
function[x,m,sd]=aaexponentialpdf(b,N)
%function[x,m,sd]=aaexponentialpdf(b,N)
for i=1:N
x(i)=-b*log(rand); %log is MATLAB function that gives the
                   %natural algorithm;
end;
m=mean(x); %mean(x)=MATLAB function providing the mean value;
sd=std(x); %std(x)=MATLAB function providing the standard
%deviation which is the square root of the variance;
```

Normal distribution

$$f(x)=\frac{1}{\sigma\sqrt{2\pi}}\exp\left(-\frac{(x-\mu)^2}{2\sigma^2}\right) \tag{3.4.7}$$

Algorithm

1. Generate two independent rv's u_1 and u_2 from uniform distribution (0,1)
2. $x_1 = (-2\ln(u_1))^{1/2} cos(2\pi u_2)$ (or $x_2 = (-2\ln u_1)^{1/2} sin(2\pi u_2)$)
3. Keep x_1 or x_2

Book MATLAB function

```
function[x]=aanormalpdf(m,s,N)
%function[x]=aanormalpdf(m,s,N);
%s=standard deviation;m=mean value;
for i=1:N
r1=rand;
r2=rand;
z(i)=sqrt(-2*log(r1))*cos(2*pi*r2);
end;
x=s*z+m;
```

Lognormal distribution

Let the rv x be $N(\mu,\sigma^2)$. Then $y = \exp(x)$ has the lognormal distribution with pdf

$$f(y)=\begin{cases}\dfrac{1}{\sqrt{2\pi}\sigma y}\exp\left[-\dfrac{(\ln y-\mu)^2}{2\sigma^2}\right] & 0\le y<\infty \\ 0 & \text{otherwise}\end{cases} \tag{3.4.8}$$

The values of σ and μ must take small values to form a lognormal-type distribution.

Algorithm

1. Generate z from N(0,1)
2. $x = \mu + \sigma z$ (x is $N(\mu,\sigma^2)$)
3. $y = \exp(x)$
4. Keep y

Book MATLAB function

```
function[y]=aalognormalpdf(m,s,N)
%function[y]=aalognormalpdf(m,s,N);
%m=mean value;s=standard deviation;N=number of samples;
for i=1:N
r1=rand;
r2=rand;
z(i)=sqrt(-2*log(r1))*cos(2*pi*r2);
end;
x=m+s*z;
y=exp(x);
```

Chi-Square distribution

If $z_{1, \ldots,} z_k$ are N(0, 1), then

$$y = \sum_{i=1}^{k} z_i^2 \tag{3.4.9}$$

has the chi-squared distribution with k degrees of freedom and is denoted by $\chi^2(k)$.

To produce graphically the pdf using any of the above functions, we may use the following MATLAB function:

```
hist(x,b);% x is a vector consisting of the values of the rv's; b is
% the number of bins required in finding the distribution of x;
```

3.5 Wiener–Khintchin relations

For a WSS process, the correlation function asymptotically goes to zero and, therefore, we can find its spectrum using the discrete-time Fourier transform. Hence, the *power spectrum* is given by

$$S_x(e^{j\omega}) = \sum_{k=-\infty}^{\infty} r_x(k)e^{-j\omega k} \tag{3.5.1}$$

This function is periodic with period 2π $(\exp(-jk(\omega + 2\pi)) = \exp(-jk\omega))$. Given the power spectral density, the autocorrelation sequence is given by the relation

$$r_x(k) = \frac{1}{2\pi}\int_{-\pi}^{\pi} S_x(e^{j\omega})e^{j\omega k}d\omega \tag{3.5.2}$$

For real process, $r_x(k) = r_x(-k)$ (symmetric function) and as a consequence the power spectrum is *even* function. Furthermore, the power spectrum of WSS process is nonnegative. These two assertions are given below in the form of mathematical relations.

$$
\begin{aligned}
&S_x(e^{j\omega}) = S_x(e^{-j\omega}) = S_x^*(e^{j\omega}) \\
&S_x(e^{j\omega}) \geq 0
\end{aligned}
\tag{3.5.3}
$$

Example 3.5.1: Find the power spectra density of the sequence $x(n) =$ *sin*(0.1*2*pi*n) + 1.5*randn(1, 32) with $n = [0 \;\; 1 \;\; 2 \ldots 31]$.

Solution: The following Book MATLAB program produced Figure 3.5.1.

```
n=0:31;
s=sin(0.1*2*pi*n);
v=randn(1,32);%Gaussian white noise;
x=s+v;
```

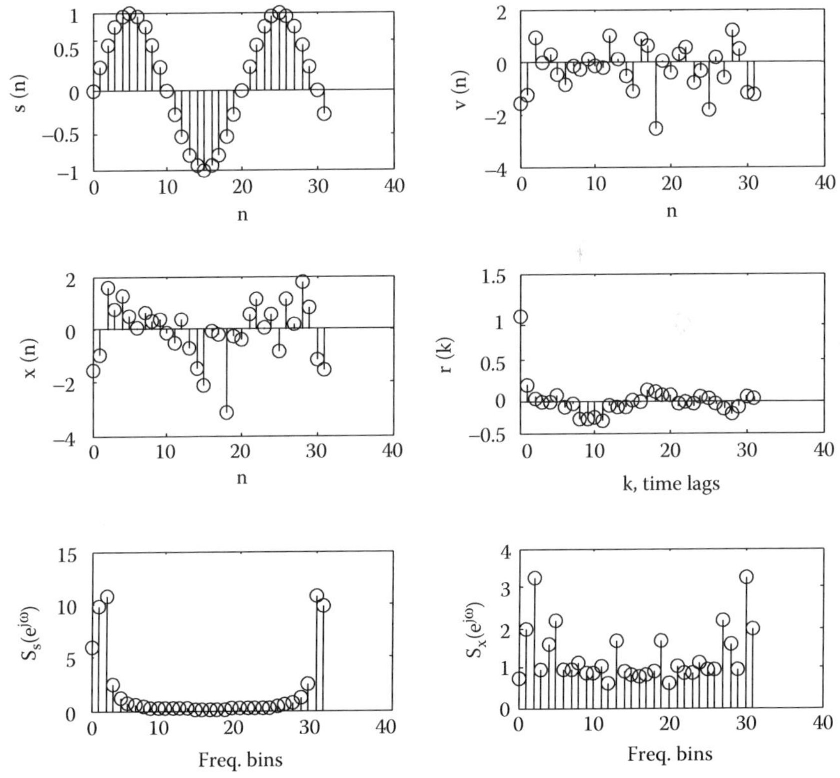

Figure 3.5.1 Illustration of Example 3.5.1.

```
r=xcorr(x,'biased');%the biased autocorrelation is divided
                    %by N, here by 32;
fs=fft(s);
fr=fft(r,32);
subplot(3,2,1);stem(n,s,'k');xlabel('n');ylabel('s(n)');
subplot(3,2,2);stem(n,v,'k');xlabel('n');ylabel('v(n)');
subplot(3,2,3);stem(n,x,'k');xlabel('n');ylabel('x(n)');
subplot(3,2,4);stem(n,r(1,32:63),'k');xlabel('k, time ...
   lags');ylabel('r(k)');
subplot(3,2,5);stem(n,abs(fs),'k');xlabel('freq. bins')...
   ;ylabel('S_s(e^{j\omega}');
subplot(3,2,6);stem(n,abs(fr),'k');xlabel('freq. bins');...
    ylabel('S_x(e^{j\omega}');
```

3.6 Filtering random processes

Linear time-invariant filters are used in many signal processing applications. Since the input signals of these filters are usually random processes, we need to determine how the statistics of these signals are modified as a result of filtering.

Let $x(n)$, $y(n)$, and $h(n)$ be the filter input, filter output, and the filter impulse response, respectively. It can be shown (see Problem 3.6.1) that if $x(n)$ is a WSS process, then the filter output autocorrelation $r_y(k)$ is related to the filter input autocorrelation $r_x(k)$ as follows.

$$r_y(k) = \sum_{l=-\infty}^{\infty} \sum_{m=-\infty}^{\infty} h(l) r_x(m-l+k) h(m) = r_x(k) * h(k) * h(-k) \tag{3.6.1}$$

The right-hand expression of (3.6.1) shows convolution of three functions. We can take the convolution of two of the functions, and the resulting function is then convolved with the third function. The results are independent of the order we operate on the functions.

From Table 2.3.1, we know that the z-transform of the convolution of two functions is equal to the product of their z-transforms. Remembering the definition of the z-transform, we find the relationship (the order of summation does not change the results)

$$Z\{h(-k)\} = \sum_{k=-\infty}^{\infty} h(-k) z^{-k} = \sum_{m=\infty}^{-\infty} h(m)(z^{-1})^{-m} = H(z^{-1}) \tag{3.6.2}$$

Therefore, the z-transform of (3.6.1) becomes

$$R_y(z) = Z\{r_x(k) * h(k)\} Z\{h(-k)\} = R_x(z) H(z) H(z^{-1}) \tag{3.6.3}$$

If we set $z = e^{j\omega}$ in the definition of the z-transform of a function, we find the spectrum of the function. Having in mind the Wiener–Khintchin theorem, (3.6.3) becomes

$$S_y(e^{j\omega}) = S_x(e^{j\omega})\left|H(e^{j\omega})\right|^2 \tag{3.6.4}$$

The above equation shows that the power spectrum of the output random sequence is equal to the power spectrum of the input sequence multiplied by the square of the absolute value of the spectrum of the filter transfer function.

Example 3.6.1: An FIR filter is defined in the time domain by the difference equation: $y(n) = x(n) + 0.5x(n-1)$. If the input signal is a white Gaussian noise, find the power spectrum of the output of the filter.

Solution: The z-transform of the difference equation is $Y(z) = (1 + 0.5z^{-1})X(z)$ (see Chapter 2). Since the ratio of the output to input is the transfer function of the filter, the transformed equation gives $H(z) = Y(z)/X(z) = 1 + 0.5z^{-1}$. The absolute value square of the spectrum of the transfer function is given by

$$\left|H(e^{j\omega})\right|^2 = (1+0.5e^{j\omega})(1+0.5e^{-j\omega}) = 1+0.5(e^{-j\omega}+e^{j\omega})+0.25 = 1.25+\cos\omega \tag{3.6.5}$$

where the Euler identity $e^{\pm j\omega} = \cos\omega \pm j\sin\omega$ was used. Figure 3.6.1 shows the sequence $x\{n\}$, the autocorrelation function $r_x\{n\}$, and the power spectrums of the filter input and its output, $S_x(\omega)$ and $S_y(\omega)$, respectively. Note that the spectrums are symmetric around $\omega = \pi$.

Spectral factorization

The power spectral density $S_x(e^{j\omega})$ of a WSS process $\{x(n)\}$ is a real-valued, positive and periodic function of ω. It can be shown that this function can be *factored* in the form

$$S_x(e^{j\omega}) = \sigma_v^2 Q(z)Q(z^{-1}) \tag{3.6.6}$$

where

1. Any WSS process $\{x(n)\}$ may be realized as the output of a causal and stable filter $h(n)$ that is driven by white noise $v(n)$ having variance σ_v^2. This is known as the *innovations representation* of the process.
2. If $x(n)$ is filtered by the filter $1/H(z)$ (*whitening filter*), the output is a white noise $v(n)$ having variance σ_v^2. This process is known as the *innovations process* and is given by

$$[\sigma_v^2 H(z)H(z^{-1})]\left[\frac{1}{H(z)H(z^{-1})}\right] = \sigma_v^2 \tag{3.6.7}$$

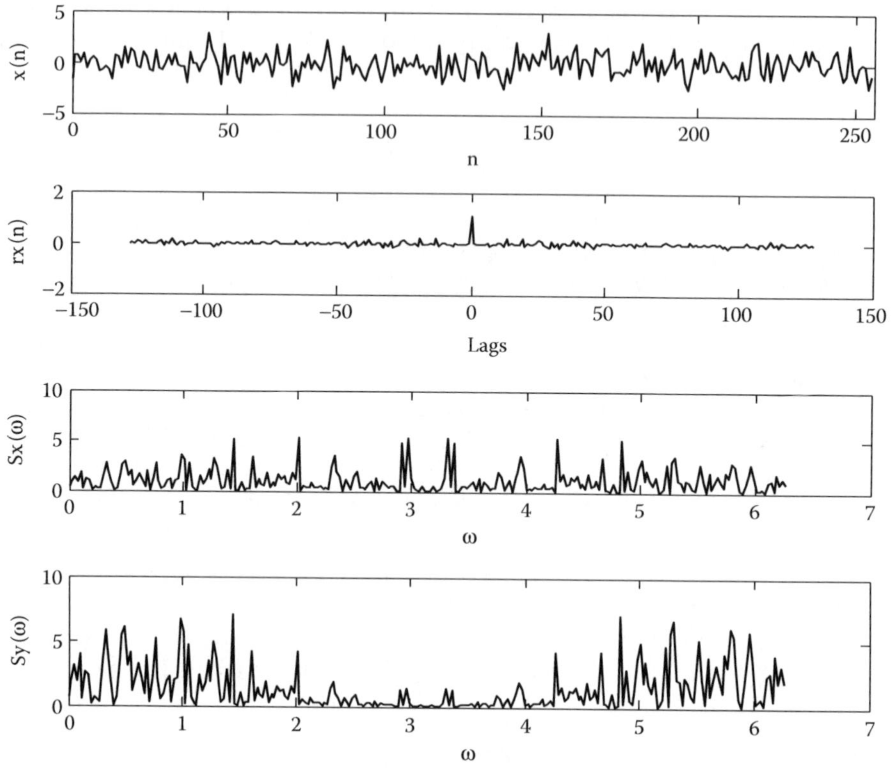

Figure 3.6.1 Illustration of Example 3.6.1.

where the first bracket is the input power spectrum, $S_x(z)$, and the second represents the filter power spectrum.

3. Since $v(n)$ and $x(n)$ are related by inverse transformations, one process can be derived from the other and they contain the same information.

3.7 Special types of random processes

Autoregressive moving average process (ARMA)

A stable shift-invariant system with p poles and q zeros can be represented in the general form

$$H(z) = \frac{B(z)}{A(z)} = \frac{\sum_{k=0}^{q} b(k)z^{-k}}{1 + \sum_{k=1}^{p} a(k)z^{-k}} \tag{3.7.1}$$

If a white noise, $v(n)$, with variance σ_v^2 is the input to the above filter, the output process $x(n)$ is a WSS process and its power spectrum is given by (see (3.6.6))

$$S_x(z) = \sigma_v^2 \frac{B(z)B(z^{-1})}{A(z)A(z^{-1})}$$

$$\text{or} \tag{3.7.2}$$

$$S_x(e^{j\omega}) = \sigma_v^2 \frac{\left|B(e^{j\omega})\right|^2}{\left|A(e^{j\omega})\right|^2}$$

This process, which has the above power spectrum density, is known as the *autoregressive moving average* process of order (p, q).

It can be shown that the following correlation relationship exists for this process:

$$r_x(k) + \sum_{m=1}^{p} a(m) r_x(k-m) = \begin{cases} \sigma_v^2 c(k) & 0 \le k \le q \\ 0 & k > q \end{cases} \tag{3.7.3}$$

$$c(k) = \sum_{m=k}^{q} b(m)h(m-k) = \sum_{m=0}^{q-k} b(m+k)h(m) \tag{3.7.4}$$

These equations are known as the *Yule–Walker equations* and can be written in the following matrix form for $k = 0, 1, \ldots, p+q$:

$$\begin{bmatrix} r_x(0) & r_x(-1) & \cdots & r_x(-p) \\ r_x(1) & r_x(0) & \cdots & r_x(-p+1) \\ \vdots & \vdots & & \vdots \\ r_x(q) & r_x(q-1) & \cdots & r_x(q-p) \\ \hline r_x(q+1) & r_x(q) & \cdots & r_x(q-p+1) \\ \vdots & \vdots & & \vdots \\ r_x(q+p) & r_x(q+p+1) & \cdots & r_x(q) \end{bmatrix} \begin{bmatrix} 1 \\ a(1) \\ a(2) \\ \vdots \\ a(p) \end{bmatrix} = \sigma_v^2 \begin{bmatrix} c(0) \\ c(1) \\ \vdots \\ c(q) \\ \hline 0 \\ \vdots \\ 0 \end{bmatrix} \tag{3.7.5}$$

It is apparent from the above equation that knowing the autocorrelation of a process produced by an ARMA model, we may be able to find the coefficients of the process. This task for the ARMA model is rather difficult due to nonlinear terms that appear in (3.7.4).

Autoregressive process (AR)

A special and important type of the ARMA process results when we set $q = 0$. Hence, from (3.7.1), (3.7.2), (3.7.3), and (3.7.4) we obtain the relations:

$$H(z) = \frac{b(0)}{1 + \sum_{k=1}^{p} a(k) z^{-k}} \tag{3.7.6}$$

$$S_x(e^{j\omega}) = \sigma_v^2 \frac{b(0)^2}{|A(e^{j\omega})|^2} \tag{3.7.7}$$

$$r_x(k) + \sum_{m=1}^{p} a(m) r_x(k-m) = \sigma_v^2 b(0)^2 \delta(k) \qquad k \geq 0 \tag{3.7.8}$$

$$\begin{bmatrix} r_x(0) & r_x(-1) & \cdots & r_x(-p) \\ r_x(1) & r_x(0) & \cdots & r_x(-p+1) \\ \vdots & \vdots & & \vdots \\ r_x(p) & r_x(p-1) & \cdots & r_x(0) \end{bmatrix} \begin{bmatrix} 1 \\ a(1) \\ \vdots \\ a(p) \end{bmatrix} = \sigma_v^2 b(0)^2 \begin{bmatrix} 1 \\ 0 \\ \vdots \\ 0 \end{bmatrix} \tag{3.7.9}$$

Example 3.7.1: Find the AR coefficients $a(1)$–$a(5)$ if the autocorrelation of the observed signal $\{x(n)\}$ is given. Assume the noise variance is equal to one. Find the power spectrum of the process.

Solution: First we produce a WSS process $\{x(n)\}$ by passing a white noise through a linear time-invariant filter. In this example we use a second order AR filter: $x(n) - 0.9x(n-1) + 0.5x(n-2) = v(n)$. The variance of the white noise input to the filter, $v(n)$, is one, and its mean value is zero. The results are shown in Figure 3.7.1, where $S(\omega) = |H(e^{j\omega})|^2$. The Book MATLAB program used to produce these results, and the AR coefficients is given below.

```
%Example 3.7.1
g=3.5*(rand(1,500)-0.5); % g: uniformly distributed WGN
                          % of zero mean and unit variance.
```

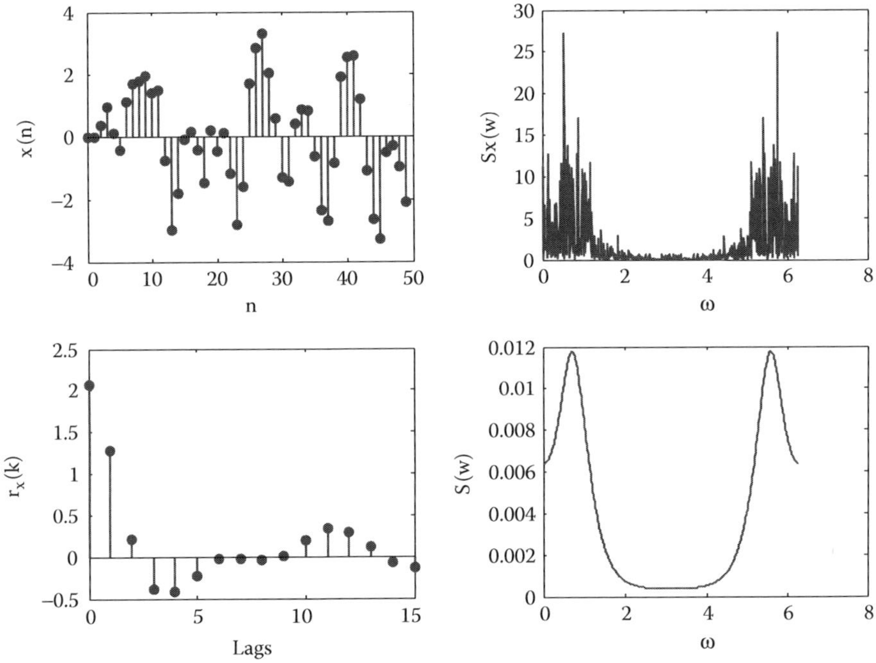

Figure 3.7.1 Illustration of Example 3.7.1.

```
x=filter(1,[1 -0.9 0.5],g);        % x: observed signal
[rx,lags]=xcorr(x,x,15,'biased'); % rx1=autocorrelation of x.
R=toeplitz(rx(1,1:6)); % we create the  6x6 autocorrelation
                       % matrix;
R1=inv(R);                        % R1=inverse of R.
b2=1/R1(1,1); b0=sqrt(b2);        % find b(0) squared.
a=b2*R1(2:6,1)                    % find the AR coefficients:
                                  % a(1) ... a(5);
H=b0./(fft([1 a'],512));          % H is the frequency response
                                  % of the filter.
nn=0:511; w=2*pi*nn/512;
S=(H.*conj(H))/512; subplot(222); plot(w,S);xlabel('\omega'); ...
    ylabel('S(w)');
subplot(221); n=0:49; stem(n,x(1:50),'filled');xlabel ...
    ('n'); ylabel('x[n]');
subplot(223); stem(lags,rx,'filled');xlabel('Lags');...
    ylabel('rx[n]');
subplot(224); X1=fft(x,512); Sx=X1.*conj(X1)/512;...
    plot(w,Sx);
xlabel('\omega');ylabel('Sx(w)');
```

Moving average process (MA)

The autocorrelation coefficient $r_x(k)$ depends nonlinearly on the filter coefficients $b(k)$, and as a consequence, estimating the coefficients is a non-trivial problem.

3.8 Nonparametric spectra estimation

The spectra estimation problem in practice is based on finite-length record $\{x(1), \ldots, x(N)\}$ of a second-order stationary random process. However, harmonic processes having line spectra appear in applications either alone or combined with noise.

Periodogram

The *periodogram spectral estimator* is based on the following formula:

$$\hat{S}_p(e^{j\omega}) = \frac{1}{N}\left|\sum_{n=0}^{N-1} x(n)e^{-j\omega n}\right|^2 = \frac{1}{N}\left|X(e^{-j\omega})\right|^2 \tag{3.8.1}$$

where $\hat{S}_p(e^{j\omega})$ is periodic with period 2π, $-\pi \le \omega \le \pi$, and $X(e^{j\omega})$ is the DFT of $x(n)$ (see Section 1.5). The periodicity is simply shown by introducing $\omega + 2\pi$ in place of ω in (3.8.1) and remembering that $\exp(j2\pi) = 1$.

Correlogram

The *correlogram spectral estimator* is based on the formula

$$\hat{S}_c(e^{j\omega}) = \sum_{m=-(N-1)}^{N-1} \hat{r}(m)e^{-j\omega m} \tag{3.8.2}$$

where $\hat{r}(m)$ is the estimate of the correlation (assumed zero mean value of $\{x(n)\}$) given by (3.2.6). It can be shown that the correlogram spectral estimator evaluated using the standard biased autocorrelation estimates coincides with that of the periodogram spectral estimator. As in (3.8.1), the correlogram is periodic function of ω with period 2π.

Computation of $\hat{S}_p(e^{j\omega})$ and $\hat{S}_p(e^{j\omega})$ using FFT

Since both functions are continuous functions of ω, we can sample the frequency as follows:

$$\omega_k = \frac{2\pi}{N}k \qquad k = 0, 1, 2, \cdots, N-1 \tag{3.8.3}$$

This situation reduces (3.8.1) and (3.8.2) to finding the DFT at those frequencies:

$$X(e^{j\omega_k}) = \sum_{n=0}^{N-1} x(n)e^{-j\frac{2\pi}{N}nk} = \sum_{n=0}^{N-1} x(n)W^{nk}, \quad 0 \le k \le N-1 \tag{3.8.4}$$

and thus,

$$\hat{S}_p(e^{j\omega_k}) = \frac{1}{N}\left|X(e^{j\omega_k})\right|^2; \quad \hat{S}_c(e^{j\omega_k}) = \sum_{m=-(N-1)}^{N-1} \hat{r}(m)e^{-j\frac{2\pi}{N}km} \tag{3.8.5}$$

The most efficient way to find the DFT using FFT is to set $N = 2^r$ for some integer r. The following two MATLAB functions give the windowed periodogram and correlogram, respectively.

Book MATLAB function for the periodogram

```
function[s]=aaperiodogram(x,w,L)
%function[s]=aaperiodogram(x,w,L)
%w=window(@name,length(x)),(name=hamming,kaiser,hann,rectwin,
%bartlett,tukeywin,blackman,gausswin,nattallwin,triang,
                    %blackmanharris);
%L=desired number of points (bins) of the spectrum;
%x=data in row form;s=complex form of the DFT;
xw=x.*w';
for m=1:L
n=1:length(x);
s(m)=sum(xw.*exp(-j*(m-1)*(2*pi/L)*n));
end;
%as=((abs(s)).^2/length(x))/norm(w)=amplitude spectral
                    %density;
%ps=(atan(imag(s)./real(s))/length(x))/norm(w)=phase spectrum;
```

To plot **as** or **ps** we can use the command: plot(0:2*pi/L:2*pi-(2*pi/L),as).

Book MATLAB function for the correlogram

```
function[s]=aacorrelogram(x,w,lg,L)
%function[s]=aacorrelogram(x,w,lg,L);
%x=data with mean zero;w=window(@name,length(2*lg)), see
                  %aaperiodogram
%function and below this function);L=desired number of
                  %spectral points;
%lg=lag number<<N;rc=symmetric autocorrelation function;
r=aasamplebiasedautoc(x,lg);
rc=[fliplr(r(1,2:lg)) r 0];
rcw=rc.*w';
for m=1:L
n=-lg+1:lg;
s(m)=sum(rcw.*exp(-j*(m-1)*(2*pi/L)*n));
end;
%asc=(abs(s).^2)/norm(w)=amplitude spectrum;
%psc=(atan(imag(s))/real(s))/norm(w)=phase spectrum;
```

General Remarks on the Periodogram

1. The variance of the periodogram does not tend to zero as $N \to \infty$. This indicates that the periodogram is an *inconsistent* estimator; that is, its distribution does not tend to cluster more closely around the true spectrum as N increases.
2. To reduce the variance and, thus, produce a smoother spectral estimator we must: a) average contiguous values of the periodogram, or b) average periodogram obtained from multiple data segments.
3. The effect of the sidelobes of the windows on the estimated spectrum consists of transferring power from strong bands to less strong bands or bands with no power. This process is known as the *leakage* problem.

Blackman–Tukey (BT) method

Because the correlation function at its extreme lag values is not reliable due to the small overlapping of the correlation process, it is recommended to use lag values about 30–40% of the total length of the data. The Blackman–Tukey estimator is a windowed correlogram and is given by

$$\hat{S}_{BT}(e^{j\omega}) = \sum_{m=-(L-1)}^{L-1} w(m)\hat{r}(m)e^{-j\omega m} \tag{3.8.6}$$

where $w(m)$ is the window with zero values for $|m| > L-1$ and $L \ll N$. The above equation can also be written in the form

$$\begin{aligned} \hat{S}_{BT}(e^{j\omega}) &= \sum_{m=-\infty}^{\infty} w(m)\hat{r}(m)e^{-j\omega m} \\ &= \hat{S}_c(e^{j\omega}) * W(e^{j\omega}) = \frac{1}{2\pi}\int_{-\pi}^{\pi} \hat{S}_c(e^{j\tau})W(e^{j(\omega-\tau)})d\tau \end{aligned} \tag{3.8.7}$$

where we applied the DTFT frequency convolution property (the DTFT of the multiplication of two functions is equal to the convolution of their Fourier transforms). Since windows have a dominant and relatively strong main lob, the BT estimator corresponds to a "locally" weighting average of the periodogram. Although the convolution smoothes the periodogram, it reduces resolution in the same time. It is expected that the smaller the L, the larger the reduction in variance and the lower the resolution. It turns out that the resolution of this spectral estimator is on the order of 1/L, whereas its variance is on the order of L/N.

For convenience, we give some of the most common windows below. For the Kaiser window the parameter β trades the main lobe width for the sidelobe leakage; $\beta = 0$ corresponds to a rectangular window, and $\beta > 0$ produces lower sidelobe at the expense of a broader main lobe.

Rectangle window

$$w(n) = 1 \qquad n = 0, 1, 2, \ldots, L-1$$

Bartlett (triangle) window

$$w(n) = \begin{cases} \dfrac{n}{\mathrm{L}/2} & n = 0, 1, \ldots, \mathrm{L}/2 \\[2ex] \dfrac{\mathrm{L}-n}{\mathrm{L}/2} & n = \dfrac{\mathrm{L}}{2}+1, \ldots, \mathrm{L}-1 \end{cases}$$

Hann window

$$w(n) = 0.5\left[1-\cos\left(\frac{2n}{\mathrm{L}}\pi\right)\right] \qquad n = 0, 1, \ldots, L-1$$

Hamming window

$$w(n) = 0.54 - 0.46\cos\left(\frac{2\pi}{\mathrm{L}}n\right) \qquad n = 0, 1, \ldots, \mathrm{L}-1$$

Blackman window

$$w(n) = 0.42 + 0.5\cos\left(\frac{2\pi}{\mathrm{L}}\left(n-\frac{\mathrm{L}}{2}\right)\right) + 0.08\cos\left(\frac{2\pi}{\mathrm{L}}2\left(n-\frac{\mathrm{L}}{2}\right)\right) \qquad n = 1, 2, \ldots, L-1$$

Kaiser window

$$w(n) = \frac{I_0\left[\beta\sqrt{1.0-\left(\dfrac{n}{\mathrm{L}/2}\right)^2}\right]}{I_0(\beta)} \qquad -(\mathrm{L}-1) \le n \le \mathrm{L}-1$$

$$I_0(x) = \sum_{k=0}^{\infty}\left[\frac{\left(\dfrac{x}{2}\right)^k}{k!}\right]^2 = \text{zero-order modified Bessel function}$$

$w(k) = 0$ for $|k| \ge \mathrm{L}$, $w(k) = w(-k)$ and equations are valid for $0 \le k \le L-1$

Note: To use the window derived from MATLAB we must write

```
w=window(@name,L)
```

```
name=the name of any of the following windows: Bartlett,
   barthannwin, blackman, blackmanharris, bohmanwin, chebwin,
   gausswin, hanning, hann, kaiser, natullwin,
   rectwin, tukeywin, triang.
L=number window values
```

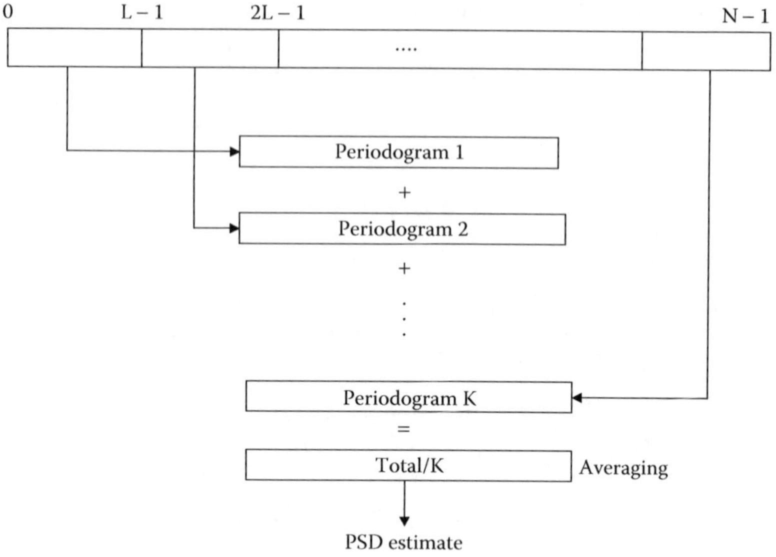

Figure 3.8.1 Bartlett method of spectra estimation.

Bartlett method

Bartlett's method reduces the fluctuation of the periodogram by splitting up the available data of N observations into $K = N/L$ subsections of L observations each, and then averaging the spectral densities of all K periodograms (see Figure 3.8.1). The MATLAB function below provides the Bartlett periodogram.

Book MATLAB function for the Bartlett method

```
function[as,ps,s]=aabartlettpsd(x,k,w,L)
%x=data;k=number of sections; w=window
                %(@name,floor(length(x)/k));
%L=number of points desired in the FT domain;
%K=number of points in each section;
K=floor(length(x)/k);
s=0;
ns=1;
for m=1:k
s=s+aaperiodogram(x(ns:ns+K-1),w,L)/k;
ns=ns+K;
end;
as=(abs(s)).^2/k;
ps=atan(imag(s)./real(s))/k;
```

Welch method

Welch proposed modifications to Bartlett method as follows: data segments are allowed to overlap and each segment is windowed prior to computing the periodogram. Since, in most practical applications, only a single realization is available, we create smaller sections as follows:

$$x_i(n) = x(iD+n)w(n) \qquad 0 \le n \le M-1, \quad 0 \le i \le K-1 \tag{3.8.8}$$

where $w(n)$ is the window of length M, D is an offset distance and K is the number of sections that the sequence $\{x(n)\}$ is divided into. Pictorially the Welch method is shown in Figure 3.8.2.

The ith periodogram is given by

$$S_i(e^{j\omega}) = \frac{1}{L}\left|\sum_{n=0}^{L-1} x_i(e^{-j\omega n})\right|^2 \tag{3.8.9}$$

and the average periodogram is given by

$$S(e^{j\omega}) = \frac{1}{K}\sum_{i=0}^{K-1} S_i(e^{j\omega}) \tag{3.8.10}$$

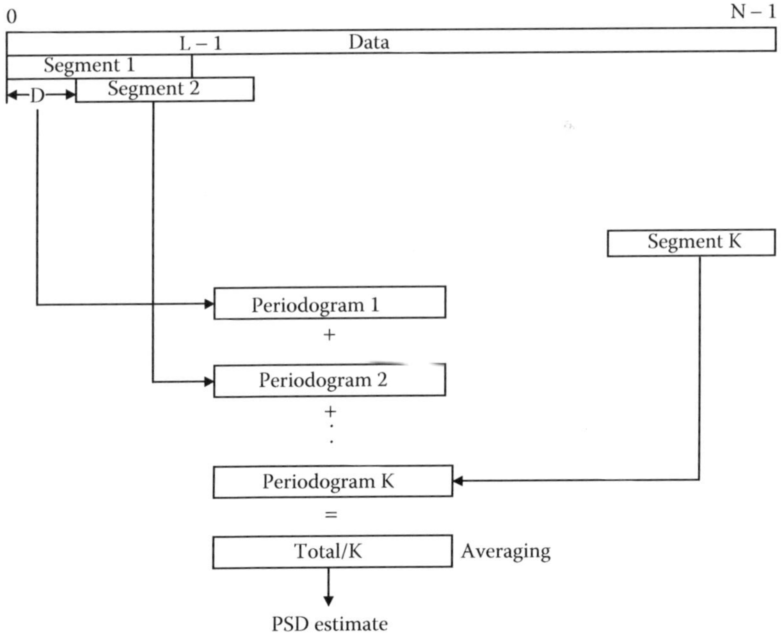

Figure 3.8.2 Welch method of spectra estimation.

If D = L, then the segments do not overlap and the result is equivalent to the Bartlett method with the exception of the data being windowed.

Book MATLAB function for the Welch method

```
function[as,ps,s,K]=aawelch(x,w,D,L)
%function[as,ps,s,K]=aawelch(x,w,D,L);
%M=length(w)=section length;
%L=number of samples desired in the frequency domain;
%w=window(@name,length of sample=length(w));x=data;
%D=offset distance=fraction of length(w),mostly 50% of
               %M;M<<N=length(x);
N=length(x);
M=length(w);
K=floor((N-M+D)/D);%K=number of processings;
s=0;
for i=1:K
s=s+aaperiodogram(x(1,(i-1)*D+1:(i-1)*D+M),w,L);
end;
as=(abs(s)).^2/(M*K);%as=amplitude spectral density;
ps=atan(imag(s)./real(s))/(M*K);%phase spectral density;
```

The MATLAB function is given as follows:

```
P=spectrum(x,m)%x=data; m=number of points of each section
%and must be a power of 2;the sections are windowed by a
%a hanning window;P is a (m/2)x2 matrix whose first column
%is the power spectral density and the second is
%the 95% confidence interval;
```

Modified Welch method

It is evident from Figure 3.8.2 that, if the lengths of the sections are not long enough, frequencies close together cannot be differentiated. Therefore, we propose a procedure, defined as *symmetric method*, and its implementation is shown in Figure 3.8.3. Windowing of the segments can also be incorporated. This approach and the rest of the proposed schemes have the advantage of progressively incorporating longer and longer segments of the data and thus introducing better and better resolution. In addition, due to the averaging process, the variance decreases and smoother periodograms are obtained. Figure 3.8.4 shows another proposed method, which is defined as the *asymmetric method*. Figure 3.8.5 shows another suggested approach for better resolution and reduced variance. The procedure is based on the method of prediction and averaging. This proposed method is defined as the *symmetric prediction method*. This procedure can be used in all the other forms, e.g., non-symmetric. The above methods can also be used for spectral estimation if we substitute the word periodogram with the word correlogram.

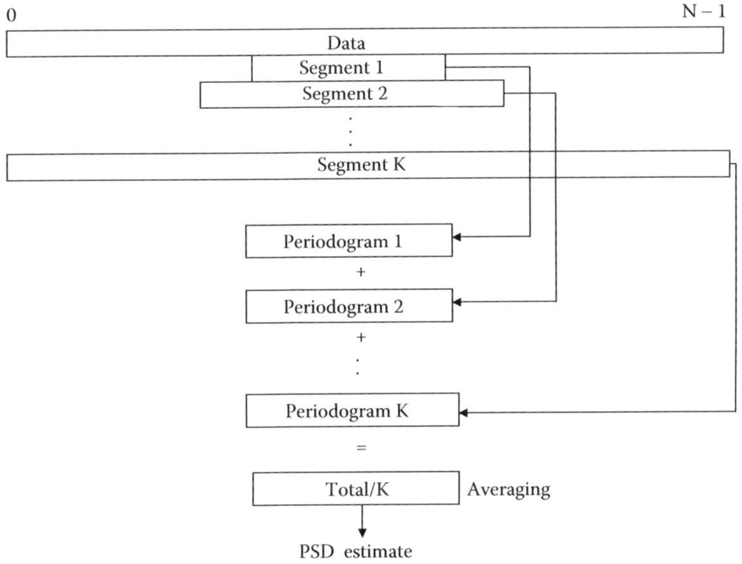

Figure 3.8.3 Modified symmetric Welch method.

Figure 3.8.6a shows data given by the equation

$$x(n) = \sin(0.3\pi n) + \sin(0.324\pi n) + 2(rand(1,128) - 0.5) \qquad (3.8.11)$$

and 128 time units. Figure 3.8.6b shows the Welch method using the MATLAB function (P=spectrum(x,64)) with the maximum length of 64 units and

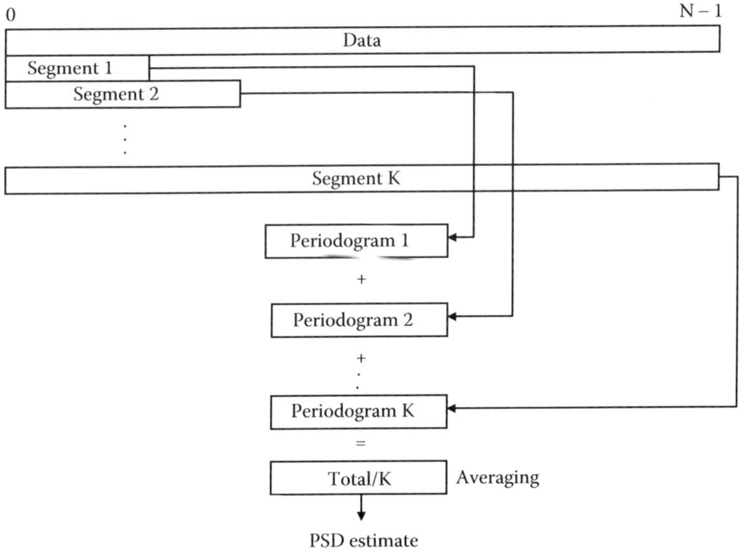

Figure 3.8.4 Modified asymmetric Welch method.

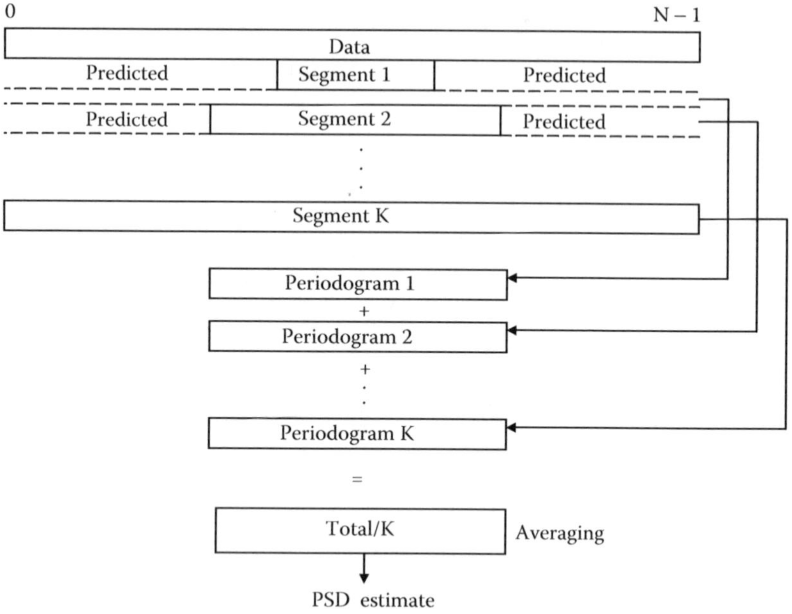

Figure 3.8.5 Modified with prediction Welch method.

windowed by a hanning window. Figure 3.8.6c shows the proposed asymmetric PSD method. It is apparent that the proposed method was successful to differentiate the two sinusoids with small frequency difference. However the variance is somewhat larger.

The Blackman–Tukey periodogram with the Bartlett window

The PSD based on the Blackman–Tukey method is given by

$$\hat{S}_{BT}(e^{j\omega}) = \sum_{m=-L}^{L} w(m)\,\hat{r}(m)e^{j\omega m}$$

$$w(m) = \begin{cases} 1 - \dfrac{|m|}{L} & m = 0, \pm 1, \pm 2, \cdots, L \\ 0 & \text{otherwise} \end{cases} \tag{3.8.12}$$

Book MATLAB function for the Blackman-Tukey periodogram with triangle window

```
function [s]=aablackmantukeypsd(x,lg,L)
%function[s]=aablackmantukeypsd(x,lg,L);
%the window used is the triangle (Bartlett) window;
%x=data;lg=lag number about 20-40% of length(x)=N;
%L=desired number of spectral points (bins);
```

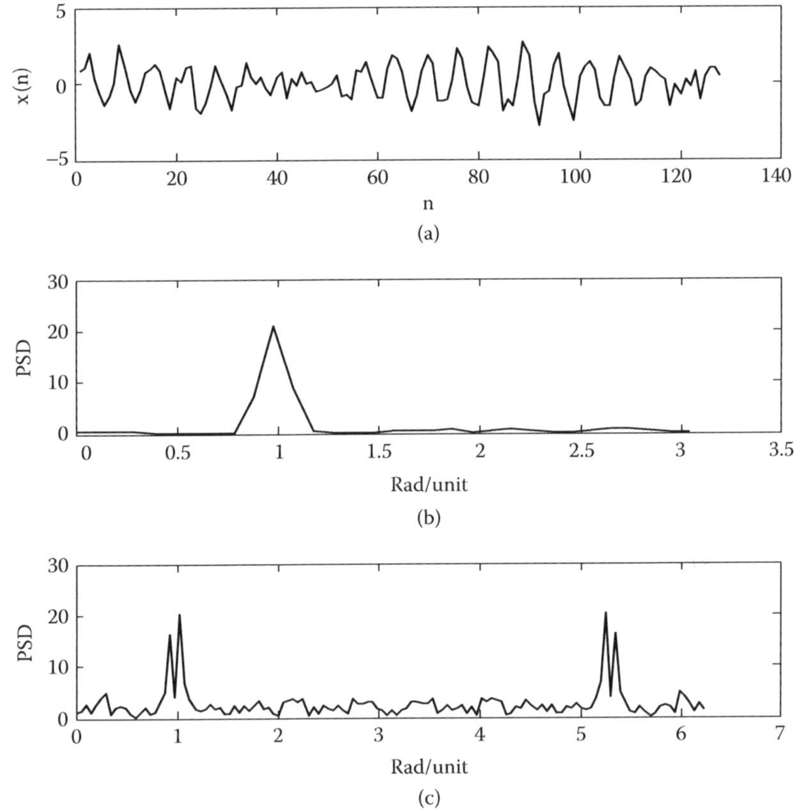

Figure 3.8.6 Comparison between Welch method (b) and modified Welch method (c).

```
[r]=aasamplediasedautoc(x,lg);
n=-(lg-1):1:(lg-1);
w=1-(abs(n)/lg);
rc=[fliplr(r(1,2:lg)) r];
rcw=rc.*w;
s=fft(rcw,L);
```

3.9 Parametric methods of power spectral estimations

The PSD of the output of a system is given by the relation (see also Section 3.7)

$$S_y(z) = H(z)H^*(1/z^*)S_v(z) = \frac{B(z)B^*(1/z^*)}{A(z)A^*(1/z^*)}S_v(z) \tag{3.9.1}$$

If $\{h(n)\}$ is a real sequence, then $H(z) = H^*(z^*)$ and (3.9.1) becomes

$$S_y = \frac{B(z)B(1/z)}{A(z)A(1/z)}S_v(z) \tag{3.9.2}$$

Since $S_v(e^{j\omega}) = \sigma_v^2$, then (3.8.1) takes the form

$$S_{yARMA}(e^{j\omega}) = \sigma_v^2 \frac{\left|B(e^{j\omega})\right|^2}{\left|A(e^{j\omega})\right|^2} = \sigma_v^2 \frac{B(e^{j\omega})B(e^{-j\omega})}{A(e^{j\omega})A(e^{-j\omega})}$$
$$= \sigma_v^2 \frac{\mathbf{e}_q^H(e^{j\omega})\mathbf{b}\mathbf{b}^H\mathbf{e}_q(e^{j\omega})}{\mathbf{e}_p^H(e^{j\omega})\mathbf{a}\mathbf{a}^H\mathbf{e}_p(e^{j\omega})} \tag{3.9.3}$$

$$\mathbf{e}_p(e^{j\omega}) = \begin{bmatrix} 1 \\ e^{j\omega} \\ \vdots \\ e^{jp\omega} \end{bmatrix}, \quad \mathbf{e}_q(e^{j\omega}) = \begin{bmatrix} 1 \\ e^{j\omega} \\ \vdots \\ e^{jq\omega} \end{bmatrix} \tag{3.9.4}$$

$$\mathbf{a} = [1\ a(1)\ a(2) \cdots a(p)]^T, \ \mathbf{b} = [b(0)\ b(1) \cdots b(q)]^T \tag{3.9.5}$$

$$A(e^{j\omega}) = 1 + \sum_{k=1}^{p} a(k)e^{-j\omega k}, \ B(e^{j\omega}) = \sum_{k=0}^{q} b(k)e^{-j\omega k} \tag{3.9.6}$$

Moving average (MA) process

Setting $a(k) = 0$ for $k = 1, 2, \ldots, p$, then

$$S_{yMA}(e^{j\omega}) = \sigma_v^2 \mathbf{e}_q^H(e^{j\omega})\mathbf{b}\mathbf{b}^H\mathbf{e}_q(e^{j\omega}), \quad \mathbf{b} = [b(0)\ b(1)\ \cdots\ b(q)]^T \tag{3.9.7}$$

Autoregressive (AR) process

Setting $b(k) = 0$ for $k = 1, 2, 3, \ldots, q$, we obtain

$$S_{yAR}(e^{j\omega}) = \sigma_v^2 \frac{1}{\mathbf{e}_p^H(e^{j\omega})\mathbf{a}\mathbf{a}^H\mathbf{e}_p(e^{j\omega})} \tag{3.9.8}$$

where $b(0)$ was set equal to one without loss of generality.

From the above development we observe that we need to find the unknown filter coefficients to be able to find the PSD of the output of the system. For the AR case the coefficients are found using the equations.

$$\mathbf{R}_{yp}\mathbf{a} = -\mathbf{r}_{yp}, \quad \mathbf{a} = -\mathbf{R}_{yp}^{-1}\mathbf{r}_{yp}$$

$$\mathbf{R}_{yp} = \begin{bmatrix} r_y(0) & r_y(-1) & \cdots & r_y(1-p) \\ r_y(1) & r_y(0) & \cdots & r_y(2-p) \\ \vdots & & & \\ r_y(p-1) & r_y(p-2) & \cdots & r_y(0) \end{bmatrix}, \quad \mathbf{a} = \begin{bmatrix} a(1) \\ a(2) \\ \vdots \\ a(p) \end{bmatrix}, \quad \mathbf{r}_{yp} = \begin{bmatrix} r_y(1) \\ r_y(2) \\ \vdots \\ r_y(p) \end{bmatrix} \tag{3.9.9}$$

Then, the variance is found from the relation

$$r_y(0) + \sum_{k=1}^{p} r(k)a(k) = \sigma_v^2 \tag{3.9.10}$$

Problems

3.1.1 If the cdf is $F(x(n)) = \begin{cases} 0 & x \leq -2 \\ 0.6 & -2 \leq x(n) < 1 \\ 1 & 1 \leq x(n) \end{cases}$, find its corresponding pdf.

3.1.2 Show that the sample mean is equal to the population mean.

3.2.1 Find the autocorrelation of the rv $x(n) = a\cos(n\omega + \theta)$, where a and ω are constants and θ is uniformly distributed over the interval $-\pi$ to π.

3.3.1 Prove the following properties of a wide-sense stationary process:

a) $\mu(n) = \mu_n = \text{constant}$
b) $r_x(-k) = r_x(k)$
c) $r_x(m,n) = r_x(m-n)$
d) $r_x(0) > r_x(k)$

3.3.2 Using Problem 3.2.1, find a 2 × 2 autocorrelation matrix.

3.4.1 Find the mean and variance of an rv having the following pdf: $f(x(n)) = (1/\sqrt{4\pi})\exp(-(x(n)-2)^2/4)$.

3.4.2 Let the rv $x(n)$ has the pdf

$$f(x(n)) = \begin{cases} \frac{1}{4}(x(n)+1) & -2 < x(n) < 2 \\ 0 & \text{otherwise} \end{cases}$$

Find the mean and variance of $x(n)$.

3.4.3 If the rv $x(n)$ is $N(\mu_n, \sigma_n^2)$, $\sigma_n^2 > 0$, then show that the rv $w_n = (x(n) - \mu_n)/\sigma_n$ is a N(0, 1).

3.4.4 The Rayleigh pdf is given by $f(x(n)) = \frac{x(n)}{\sigma_n^2} e^{-x^2(n)/2\sigma_n^2} u(x(n))$, where $u(x(n))$ is the unit step function.

a) Plot $f(x(n))$ by changing the parameter σ_n.
b) Determine the mean and variance.

Hints-solutions-suggestions

3.1.1:

$$f(x) = 0.6\delta(x+2) + 0.4\delta(x-1)$$

3.1.2:

$$E\{\hat{\mu}\} = E\left\{\frac{1}{N}\sum_{n=0}^{N-1} x(n)\right\} = \frac{1}{N}\sum_{n=0}^{N-1} E\{x(n)\} = \frac{1}{N}\sum_{n=0}^{N-1} \mu = \mu$$

3.2.1:

$$\mu_n = E\{x(n)\} = E\{a\sin(n\omega + \theta)\} = a\int_{-\pi}^{\pi} \sin(n\omega + \theta)\frac{1}{2\pi}d\theta = 0.$$

$$r_x(m,n) = E\{a^2 \sin(m\omega + \theta)\sin(n\omega + \theta)\} = (1/2)a^2 E\{\cos[(m-n)\omega] - (1/2)$$

$$\times \cos[(m+n)\omega + 2\theta]\} = (1/2)a^2 \cos[(m-n)\omega]$$

because the ensemble of the cosine with theta is zero and the other cosine is a constant independent of the rv theta.

3.3.1:

a) $E\{x(n+q)\} = E\{x(m+q)\}$ implies that that the mean must be a constant.

b) $r_x(k) = E\{x(n+k)x(n)\} = E\{x(n)x(n+k)\} = r_x(n-n-k) = r_x(-k)$

c) $r_x(m,n) = \int_{-\infty}^{\infty}\int_{-\infty}^{\infty} x(m)x(n) f(x(m+k)x(n+k))dx(m)dx(n) = r_x(m+k, n+k)$

$$= r_x(m-n, 0) = r_x(m-n)$$

d) $E\{[x(n+k) - x(n)]^2\} = r_x(0) - 2r_x(k) + r_x(0) \geq 0$ *or* $r_x(0) \geq r_x(k)$.

3.3.2:

$$\mathbf{R}_x = (a^2/2)\begin{bmatrix} 1 & \cos\omega \\ \cos\omega & 1 \end{bmatrix}$$

3.4.1:

$$\mu_n = \int_{-\infty}^{\infty} x \frac{1}{\sqrt{4\pi}} e^{-(x-2)^2/4} dx, \text{ set } x-2=y \Rightarrow \mu_n = \frac{1}{\sqrt{4\pi}} \int_{-\infty}^{\infty} (y+2) e^{-y^2/4} dy$$

$$= 0 + \frac{1}{\sqrt{\pi}} 2\sqrt{\pi} = 2$$

since y is an odd function and the first integral vanishes. The second integral is found using tables.

$$\sigma^2 = \int_{-\infty}^{\infty} (x-2)^2 \frac{1}{\sqrt{4\pi}} e^{-(x-2)^2/4} dx$$

$$= \frac{1}{\sqrt{4\pi}} \int_{-\infty}^{\infty} y^2 e^{-y^2/4} dy = \frac{1}{\sqrt{4\pi}} \frac{1}{2(1/4)} \sqrt{4\pi} = 2 \text{ (using tables)}$$

3.4.2:

$$\mu_n = \int_{-\infty}^{\infty} x(n) f(x(n)) dx(n) = \int_{-2}^{2} x(n)(1/4)(x(n)+1) dx(n) = \frac{1}{4}\left[\frac{x^3(n)}{3} + \frac{x^2(n)}{2}\right]\Bigg|_{-2}^{2} = 1,$$

$$\sigma_n^2 = \int_{-\infty}^{\infty} x^2(n) \frac{1}{4}(x(n)+1) dx(n) = 1$$

3.4.3:

$$W(w_n) = cdf = \Pr\{\frac{x(n)-\mu_n}{\sigma_n} \le w_n\} = \Pr\{x(n) \le w_n\sigma_n + \mu_n\} \quad \Rightarrow$$

$$W(w_n) = \int_{-\infty}^{w_n\sigma_n+\mu_n} \frac{1}{\sigma_n\sqrt{2\pi}} \exp(-\frac{(x(n)-\mu_n)^2}{2\sigma_n^2}) dx(n).$$

Change variables $y_n = (x(n)-\mu_n)/\sigma_n \quad \Rightarrow$

$$W(w_n) = \int_{-\infty}^{w_n} \frac{1}{\sqrt{2\pi}} e^{-y_n^2/2} dy_n . But \ f(w_n) = \frac{dW(w_n)}{dw_n} \quad \Rightarrow$$

$$f(w_n) = \frac{1}{\sqrt{2\pi}} \exp(-w_n^2/2) \quad -\infty < w_n < \infty \quad \Rightarrow$$

w_n is $N(0,1)$

3.4.4:

$$b)\,\mu_n=\int_{-\infty}^{\infty}x(n)f(x(n))dx(n)=\int_0^{\infty}\frac{x^2(n)}{\sigma_n^2}\exp(-x^2(n)/2\sigma^2)dx(n)$$

$$=-\int_0^{\infty}x(n)d(e^{-x^2(n)/2\sigma^2})$$

$$=-x(n)\exp(-x^2(n)/2\sigma_n^2)\Big|_0^{\infty}+\int_0^{\infty}\exp(-x^2(n)/2\sigma_n)dx(n)$$

$$=\frac{\sqrt{\pi}}{2\sqrt{1/2\sigma_n^2}}=\sigma_n\sqrt{\frac{\pi}{2}}$$

$$\sigma_n^2=\int_0^{\infty}(x(n)-\sigma_n\sqrt{\frac{\pi}{2}})^2\frac{x(n)}{\sigma_n^2}\exp(-x^2(n)/2\sigma_n^2)dx(n)$$

$$=-\int_0^{\infty}(x(n)-\sigma_n\sqrt{\frac{\pi}{2}})^2 d(\exp(-x^2(n)/2\sigma_n^2))$$

$$=-(x(n)-\sigma_n\sqrt{\pi/2})^2\exp(-x(n)/2\sigma_n^2)\Big|_0^{\infty}$$

$$+\int_0^{\infty}2(x(n)-\sigma_n^2\sqrt{\pi/2})\exp(-x^2(n)/2\sigma_n^2)dx(n)$$

$$=-\sigma_n^2(\pi/2)-\sigma_n^2\pi-\sigma_n^2\int_0^{\infty}d(\exp(-x^2(n)/2\sigma_n^2)=\sigma_n^2\frac{\pi}{2}+\sigma_n^2$$

chapter 4

Wiener filters

4.1 The mean-square error

In this chapter we develop a class of linear optimum discrete-time filters known as the *Wiener filters*. These filters are optimum in the sense of minimizing an appropriate function of the error, known as the *cost function*. The cost function that is commonly used in filter design optimization is the *mean-square error* (MSE). Minimizing MSE involves only second-order statistics (correlations) and leads to a theory of linear filtering that is useful in many practical applications. This approach is common to all optimum filter designs. Figure 4.1.1 shows the block diagram presentation of the optimum filter problem.

The basic idea is to recover a desired signal $d(n)$ given a noisy observation $x(n) = d(n) + v(n)$, where both $d(n)$ and $v(n)$ are assumed to be WSS processes. Therefore, the problem can be stated as follows:

Design a filter that produces an estimate $\hat{d}(n)$ of the desired signal $d(n)$ using a linear combination of the data $x(n)$ such that the MSE function (cost function)

$$J = E\{(d(n) - \hat{d}(n))^2\} = E\{e^2(n)\} \tag{4.1.1}$$

is minimized.

Depending on how the data $x(n)$ and the desired signal $d(n)$ are related, there are four basic problems that need solutions. These are: Filtering, Smoothing, Prediction, and Deconvolution.

4.2 The FIR Wiener filter

Let the sample response (filter coefficients) of the desired filter be denoted by $\mathbf{w}$. This filter will process the real-valued stationary process $\{x(n)\}$ to produce an estimate $\hat{d}(n)$ of the desired real-valued signal $d(n)$. Without loss of generality, we will assume, unless otherwise stated, that the processes $\{x(n)\}$, $\{d(n)\}$, etc., have zero mean values. Furthermore, assuming that the

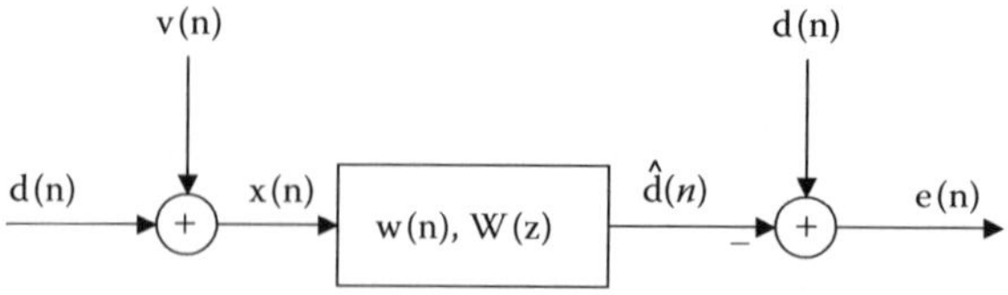

Figure 4.1.1 Block diagram of the optimum filtering problem.

filter coefficients do not change with time, the output of the filter is equal to the convolution of the input and the filter coefficients. Hence, we obtain

$$\hat{d}(n) = \sum_{m=0}^{M-1} w_m x(n-m) = \mathbf{w}^{\mathrm{T}}\mathbf{x}(n) \tag{4.2.1}$$

where M is the number of filter coefficients, and

$$\mathbf{w} = [w_1 \ w_2 \ \cdots \ w_{M-1}]^{\mathrm{T}}, \ \mathbf{x}(n) = [x(n) \ x(n-1) \ \cdots \ x(n-M+1)]^{\mathrm{T}} \tag{4.2.2}$$

The MSE is given by (see (4.1.1))

$$\begin{aligned} J(\mathbf{w}) &= E\{e^2(n)\} = E\{[d(n) - \mathbf{w}^{\mathrm{T}}\mathbf{x}(n)]^2\} \\ &= E\{[d(n) - \mathbf{w}^{\mathrm{T}}\mathbf{x}(n)][d(n) - \mathbf{w}^{\mathrm{T}}\mathbf{x}(n)]^{\mathrm{T}} \\ &= E\{d^2(n) - \mathbf{w}^{\mathrm{T}}\mathbf{x}(n)d(n) - d(n)\mathbf{x}^{\mathrm{T}}(n)\mathbf{w} + \mathbf{w}^{\mathrm{T}}\mathbf{x}(n)\mathbf{x}^{\mathrm{T}}(n)\mathbf{w}\} \\ &= E\{d^2(n)\} - 2\mathbf{w}^{\mathrm{T}}E\{d(n)\mathbf{x}(n)\} + \mathbf{w}^{\mathrm{T}}E\{\mathbf{x}(n)\mathbf{x}^{\mathrm{T}}(n)\}\mathbf{w} \\ &= \sigma_d^2 - 2\mathbf{w}^{\mathrm{T}}\mathbf{p}_{dx} + \mathbf{w}^{\mathrm{T}}\mathbf{R}_x\mathbf{w} \end{aligned} \tag{4.2.3}$$

where

$$\begin{aligned} &\mathbf{w}^{\mathrm{T}}\mathbf{x}(n) = \mathbf{x}^{\mathrm{T}}(n)\mathbf{w} = \text{scalar} \\ &\sigma_d^2 = \text{variance of the desired signal, d(n)} \\ &\mathbf{p}_{dx} = [p_{dx}(0) \ p_{dx}(1) \ \cdots \ p_{dx}(M-1)]^{\mathrm{T}} \\ &\quad = \text{cross-correlation vector between d(n) and } \mathbf{x}(n) \\ &p_{dx}(0) = r_{dx}(0), \ p_{dx}(1) = r_{dx}(1), \ \cdots \ , p_{dx}(M-1) = r_{dx}(M-1) \end{aligned} \tag{4.2.4}$$

$$\mathbf{R}_x = E\left\{\begin{bmatrix} x(n) \\ x(n-1) \\ \vdots \\ x(n-M+1) \end{bmatrix} [x(n) \ x(n-1) \ \cdots \ x(n-M+1)]\right\}$$

$$= \begin{bmatrix} E\{x(n)x(n)\} & E\{x(n)x(n-1)\} & \cdots & E\{x(n)x(n-M+1)\} \\ E\{x(n-1)x(n)\} & E\{x(n-1)x(n-1)\} & \cdots & E\{x(n-1)x(n-M+1)\} \\ \vdots & & & \vdots \\ E\{x(n-M+1)x(n)\} & E\{x(n-M+1)x(n-1)\} & \cdots & E(x(n-M+1)x(n-M+1)\} \end{bmatrix}$$

$$= \begin{bmatrix} r_x(0) & r_x(1) & \cdots & r_x(M-1) \\ r_x(-1) & r_x(0) & \cdots & r_x(M-2) \\ \vdots & & & \vdots \\ r_x(-M+1) & r_x(-M+2) & \cdots & r_x(0) \end{bmatrix} \tag{4.2.5}$$

The above matrix is the correlation matrix of the input data, and it is symmetric because the random process is assumed to be stationary, and hence, we have the equality, $r_x(k) = r_x(-k)$. Since in practical cases we have only one realization, we will assume that the signal is ergodic. Therefore, we will use the sample autocorrelation coefficients given in Section 3.2.

Example 4.2.1: Let us assume that we have found the sample autocorrelation coefficients ($r_x(0) = 1.0$, $r_x(1) = 0$) from given data $x(n)$, which, in addition to noise, contain the desired signal. Furthermore, let the variance of the desired signal $\sigma_d^2 = 24.40$ and the cross-correlation vector be $\mathbf{p}_{dx} = [2 \ \ 4.5]^T$. It is desired to find the surface defined by the mean-square function $J(\mathbf{w})$.

Solution: Substituting the values given above in (4.2.3), we obtain

$$J(\mathbf{w}) = 24.40 - 2[w_0 \ \ w_1]\begin{bmatrix} 2 \\ 4.2 \end{bmatrix} + [w_0 \ \ w_1]\begin{bmatrix} 1 & 0 \\ 0 & 1 \end{bmatrix}\begin{bmatrix} w_0 \\ w_1 \end{bmatrix} \tag{4.2.6}$$

$$= 24.40 - 4w_0 - 9w_1 + w_0^2 + w_1^2$$

Note that the equation is quadratic with respect to filter coefficients, and it is true for any number of filter coefficients. This is because we used the mean-square error approach for the minimization of the error. Figure 4.2.1 shows the schematic representation of the Wiener filter. The data are the sum of the desired signal and noise. From the data we find the correlation matrix

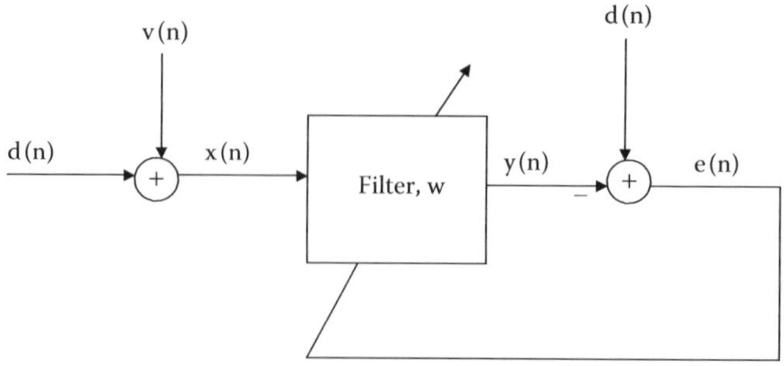

Figure 4.2.1 A schematic presentation of Wiener filtering.

and the cross-correlation between the desired signal and the data. Note that to find the optimum Wiener filter coefficients, the desired signal is needed. Figure 4.2.2 shows the *MSE surface*. This surface is found by inserting different values of w_0 and w_1 in the function $J(\mathbf{w})$. The values of the coefficients that correspond to the bottom of the surface are the *optimum* Wiener coefficients. The vertical distance from the w_0-w_1 plane to the bottom of the surface is known as the *minimum error*, J_{min}, and corresponds to the optimum Wiener coefficients. We observe that the minimum height of the surface corresponds to about $w_0 = 2$ and $w_1 = 4.5$, which are the optimum coefficients, as we will learn how to find them in the next section. Figure 4.2.3 shows an adaptive FIR filter.

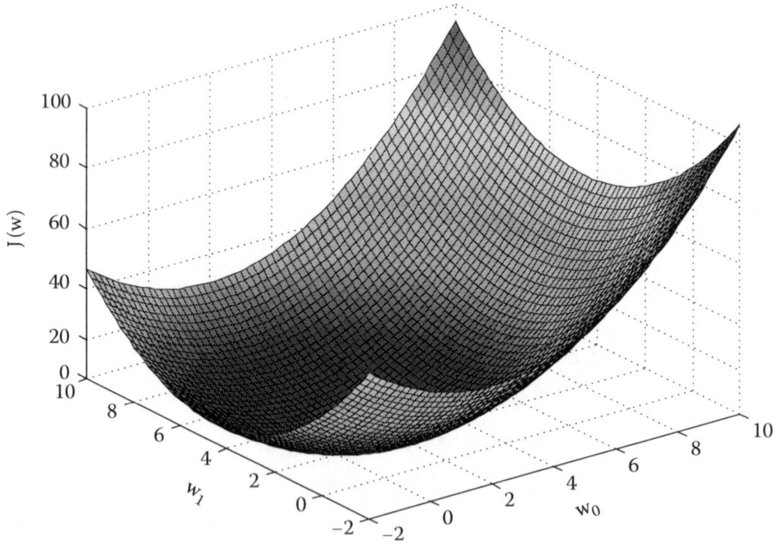

Figure 4.2.2 The mean-square error surface.

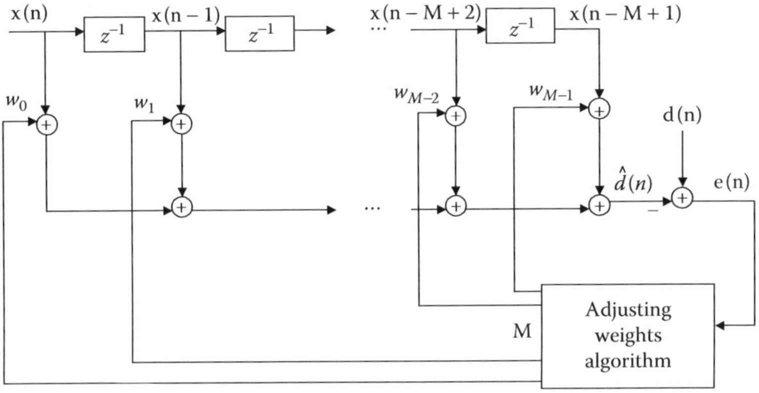

Figure 4.2.3 An adaptive FIR filter.

4.3 The Wiener solution

From Figure 4.2.2, we observe that there exists a plane touching the parabolic surface at its minimum point, and it is parallel to the **w**-plane. Furthermore, we observe that the surface is concave upward, and therefore, the first derivative of the MSE with respect to w_0 and w_1 must be zero at the minimum point, and the second derivative must be positive. Hence, we write

$$\frac{\partial J(w_0, w_1)}{\partial w_0} = 0 \qquad \frac{\partial J(w_0, w_1)}{\partial w_1} = 0 \qquad \text{(a)}$$
$$\frac{\partial^2 (w_0, w_1)}{\partial^2 w_0} > 0 \qquad \frac{\partial^2 J(w_0, w_1)}{\partial^2 (w_1)} > 0 \qquad \text{(b)} \qquad (4.3.1)$$

For a two-coefficient filter, (4.2.3) becomes

$$J(w_0, w_1) = w_0^2 r_x(0) + 2w_0 w_1 r_x(1) + w_1^2 r_x(0) - 2w_0 r_{dx}(0) - 2w_1 r_{dx}(1) + \sigma_d^2$$
$$(4.3.2)$$

Introducing (4.3.2) in part (a) of (4.3.1) produces the following set of equations

$$2w_0^o r_x(0) + 2w_1^o r_x(1) - 2r_{dx}(0) = 0 \qquad \text{(a)}$$
$$2w_1^o r_x(0) + 2w_0^o r_x(1) - 2r_{dx}(1) = 0 \qquad \text{(b)} \qquad (4.3.3)$$

The above system can be written in the following matrix form, called the Wiener–Hopf equation:

$$\mathbf{R}_x \mathbf{w}^o = \mathbf{p}_{dx} \tag{4.3.4}$$

where the superscript "o" indicates the optimum Wiener solution for the filter. Note that to find the correlation matrix we must know the second-order statistics. If, in addition, the matrix is invertible, which is the case in most practical signal processing applications, the optimum filter is given by

$$\mathbf{w}^o = \mathbf{R}_x^{-1} \mathbf{p}_{dx} \tag{4.3.5}$$

For an M-order filter, $\mathbf{R}_x$ is an $M \times M$ matrix, $\mathbf{w}^o$ is an $M \times 1$ vector, and $\mathbf{p}$ is an $M \times 1$ vector.

If we differentiate (4.3.3) once more with respect to w_0^o and w_1^o (i.e., differentiating $J(\mathbf{w})$ twice), we find that the result is equal to $2r_x(0)$. Since $r_x(0) = E\{x(m)x(m)\} = \sigma_x^2 > 0$, the surface is concave upward. Therefore, the extreme is the minimum point of the surface. Furthermore, if we substitute (4.3.5) in (4.2.3), we obtain the minimum error in the mean-square sense (see Problem 4.3.1)

$$J_{\min} = \sigma_d^2 - \mathbf{p}_{dx}^T \mathbf{w}^o = \sigma_d^2 - \mathbf{p}_{dx}^T \mathbf{R}_x^{-1} \mathbf{p}_{dx} \tag{4.3.6}$$

which indicates that the minimum point of the error surface is at a distance $J_{\min}$ above the $\mathbf{w}$-plane. The above equation shows that if no correlation exists between the desired signal and the data, the error is equal to the variance of the desired signal.

The problem we are facing is how to choose the length of the filter M. In the absence of *a priori* information, we compute the optimum coefficients, starting from a small reasonable number. As we increase the number, we check the MMSE, and if its value is small enough, e.g., MMSE < 0.01, we accept the corresponding number of the coefficients.

Example 4.3.1: We would like to find the optimum filter coefficients w_0 and w_1 of the Wiener filter, which approximates (models) the unknown system with coefficients $b_0 = 1$ and $b_1 = 0.38$ (see Figure 4.3.1).

Solution: The following Book MATLAB program was used:

```
% example4_3_1 m-file

v =0.5*(rand(1,20)-0.5);%v=noise vector(20 uniformly
%distributed rv's with mean zero);
x=randn(1,20);% x=data vector entering the system and
%the Wiener filter (20 normal
%distributed rv's with mean zero);
```

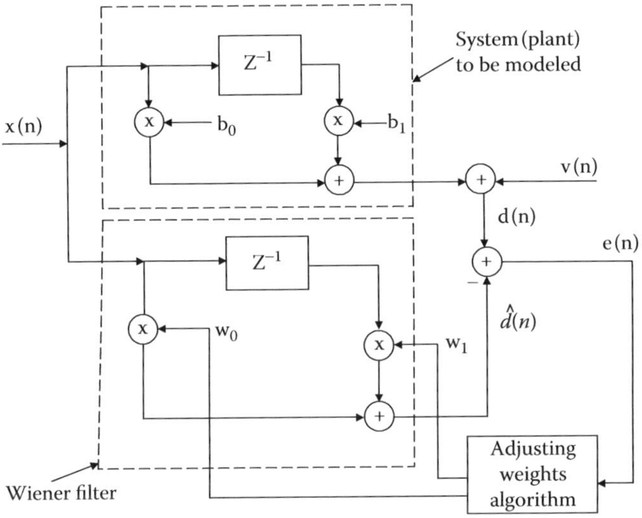

Figure 4.3.1 Illustration of Example 4.3.1.

```
sysout=filter([1 0.38],1,x);% sysout=system output
%with x as input; filter(b,a,x) is a
%MATLAB function, where
%b is the vector of the coefficients of the ARMA
%numerator, a is the vector of
%the coefficients of the ARMA denominator (see (2.4.7);
dn=sysout +v;
rx=aasamplebiasedautoc(x,2);%book MATLAB function with
%lag=2;
Rx=toeplitz(rx);%toeplitz() is a MATLAB function that
%gives the symmetric autocorrelation matrix;
pdx=xcorr(x,dn,'biased');%xcorr() a MATLAB function
%that gives a symmetric biased crosscorrelation;
p=pdx(1,19:20);
w=inv(Rx)*p';
dnc=aasamplebiasedautoc(dn,1);% s2dn=variance of the
%desired signal;
jmin=dnc-p*w;
```

Some representative values found in this example are: $\mathbf{R}_x$= [0.9778 0.0229 0.0229 0.9675], $\mathbf{p}$ = [0.3916 1.0050], $\mathbf{w}$ = [1.0190 0.3767]; J_{min} = 0.0114. We observe that the Wiener filter coefficients are close to those of the unknown system.

Orthogonality condition

In order for the set of filter coefficients to minimize the cost function $J(\mathbf{w})$, it is necessary and sufficient that the derivatives of $J(\mathbf{w})$ with respect to w_k be equal to zero for $k = 0, 1, 2, \ldots, M-1$,

$$\frac{\partial J}{\partial w_k} = \frac{\partial}{\partial w_k} E\{e(n)e(n)\} = 2E\{e(n)\frac{\partial e(n)}{\partial w_k}\} = 0 \tag{4.3.7}$$

But

$$e(n) = d(n) - \sum_{m=0}^{M-1} w_m x(n-m) \tag{4.3.8}$$

and, hence, it follows that

$$\frac{\partial e(n)}{\partial w_k} = -x(n-k) \tag{4.3.9}$$

Therefore, (4.3.7) becomes

$$E\{e^o(n)x(n-k)\} = 0 \qquad k = 0, 1, 2, \ldots, M-1 \tag{4.3.10}$$

where the superscript "o" denotes that the corresponding w_k's used to find the estimation error $e^o(n)$ are the optimal ones. Figure 4.3.2 illustrates the orthogonality principle, where the error $e^o(n)$ is orthogonal (perpendicular) to the data set $\{x(n)\}$ when the estimator employs the optimum set of filter coefficients.

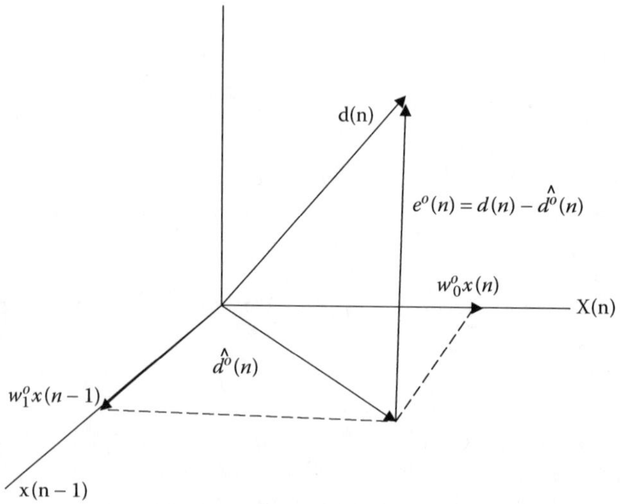

Figure 4.3.2 Pictorial illustration of the orthogonality principle.

4.4 Wiener filtering examples

The examples in this section will illustrate the use and utility of the Wiener filters.

Example 4.4.1 (Filtering): Filtering of noisy signals (noise reduction) is extremely important, and the method has been used in many applications, such as speech in noisy environment, reception of data across a noisy channel, enhancement of images, etc.

Let the received signal be $x(n) = d(n) + v(n)$, where $v(n)$ is a noise with zero mean, variance σ_v^2, and it is uncorrelated with the desired signal, $d(n)$. Hence,

$$\begin{aligned} p_{dx}(m) &= E\{d(n)x(n-m)\} = E\{d(n)d(n-m)\} + E\{d(n)\}E\{v(n-m)\} \\ &= E\{d^2(m)\} = r_d(m) \end{aligned} \tag{4.4.1}$$

Similarly, we obtain

$$r_x(m) = E\{x(n)x(n-m)\} = r_d(m) + r_v(m) \tag{4.4.2}$$

where we used the assumption that $d(n)$ and $v(n)$ are uncorrelated, and $v(n)$ has zero mean value. Therefore, the Wiener-Hopf equation (Equation 4.3.4) becomes

$$(\mathbf{R_d} + \mathbf{R_v})\mathbf{w}^o = \mathbf{p_{dx}} \tag{4.4.3}$$

The following Book MATLAB program was used to produce the results shown in Figure 4.4.1.

```
%example4_4_1  m-file
n=0:511;
d=sin(.1*pi*n);%desired signal
v=0.5*randn(1,512);%white Gaussian noise;
x=d+v;%input signal to Wiener filter;
rd=aasamplebiasedautoc(d,20);%rdx=rd=biased autoc.
%function of the desired signal(see 4.4.1);
rv=aasamplebiasedautoc(v,20);%rv=biased autoc. func-
%tion of the noise;
R=toeplitz(rd(1,1:12))+toeplitz(rv(1,1:12));%see(4.4.
%3);
pdx=rd(1,1:12);
w=inv(R)*pdx';
y=filter(w',1,x);%output of the filter;
```

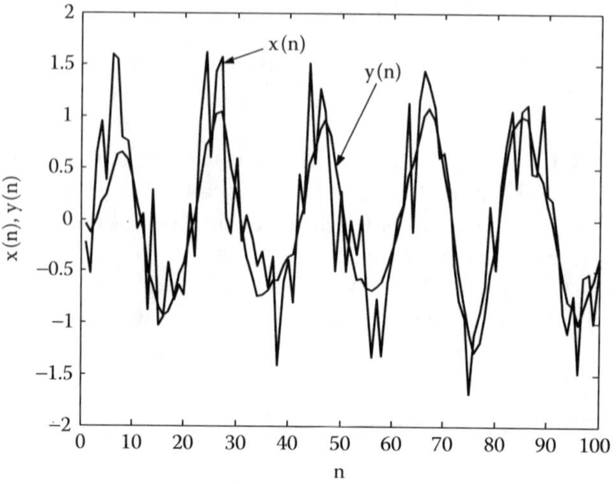

Figure 4.4.1 Illustration of Example 4.4.1.

But

$$\sigma_x^2 = \sigma_d^2 + \sigma_v^2; \quad (\sigma_x^2 = r_x(0),\ \sigma_d^2 = r_d(0),\ \sigma_v^2 = r_v(0)) \tag{4.4.4}$$

and, hence, from the MATLAB function *var* we obtained var(d) = var(x) – var(v) = 0.4968 and $J_{min} = 0.4968 - \mathbf{p}_{dx}\mathbf{w}^o = 0.0320$.

We can also use the Book MATLAB function [w,jmin] = aawienerfirfilter(x,d,M), to obtain the filter coefficients and the minimum MSE.

Example 4.4.2 (Filtering): It is desired to find a two-coefficient Wiener filter for the communication channel shown in Figure 4.4.2. Let $v_1(n)$ and $v_2(n)$ are white noises with zero mean, uncorrelated with each other and with $d(n)$, and have the following variances: $\sigma_1^2 = 0.31$, $\sigma_2^2 = 0.12$. The desired signal produced by the first filter shown in Figure 4.4.2 is

$$d(n) = -0.796d(n-1) + v_1(n) \tag{4.4.5}$$

Therefore, the autocorrelation function of the desired signal becomes

$$E\{d(n)d(n)\} = 0.796^2 E\{d^2(n-1)\} - 2\times 0.796E\{d(n-1)\}E\{v_1(n)\} + E\{v_1^2(n)\}$$

or

$$\sigma_d^2 = 0.796^2\sigma_d^2 + \sigma_1^2 \quad \text{or} \quad \sigma_d^2 = 0.31/(1-0.796^2) = 0.8461 \tag{4.4.6}$$

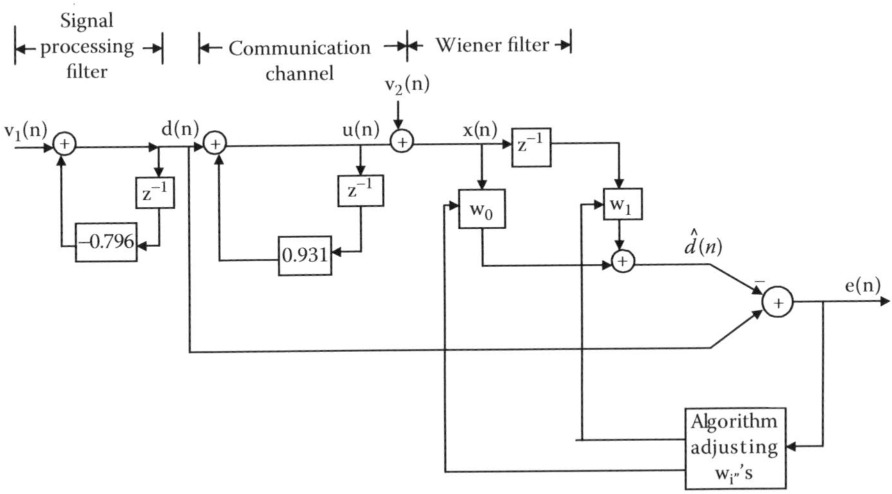

Figure 4.4.2 Illustration of Example 4.4.2.

From the second filter we obtain

$$d(n) = u(n) - 0.931u(n-1) \tag{4.4.7}$$

Introducing (4.4.7) in (4.4.5) we obtain

$$u(n) - 0.135u(n-1) - 0.7411u(n-2) = v_1(n) \tag{4.4.8}$$

But $x(n) = u(n) + v_2(n)$ and, hence, the vector form of the set becomes $\mathbf{x}(n) = \mathbf{u}(n) + \mathbf{v}_2(n)$. Therefore, the autocorrelation matrix $\mathbf{R}_x$ becomes

$$E\{\mathbf{x}(n)\mathbf{x}^T(n)\} = \mathbf{R}_x = E\{[\mathbf{u}(n) + \mathbf{v}_2(n)][\mathbf{u}^T(n) + \mathbf{v}_2^T(n)]\} = \mathbf{R}_u + \mathbf{R}_{v_2} \tag{4.4.9}$$

where we used the assumption that $\mathbf{u}(n)$ and $\mathbf{v}_2(n)$ are uncorrelated zero-mean random sequences, which implies that $E\{\mathbf{v}_2(n)\mathbf{u}^T(n)\} = E\{\mathbf{v}_2(n)\}E\{\mathbf{u}(n)\} = 0$.

Next, we multiply (4.4.8) by $u(n-m)$ and take the ensemble average of both sides, which results in the following expression

$$r_u(m) - 0.135r_u(m-1) - 0.7411r_u(m-2) = r_{v_1u}(m) = E\{v_1(n)u(n-m)\} \tag{4.4.10}$$

Setting $m = 1$ and $m = 2$, the above equation produces the following system:

$$\begin{bmatrix} r_u(0) & r_u(-1) \\ r_u(1) & r_u(0) \end{bmatrix} \begin{bmatrix} -0.1350 \\ -0.7411 \end{bmatrix} = \begin{bmatrix} -r_u(1) \\ -r_u(2) \end{bmatrix} = \text{Yule–Walker equation} \tag{4.4.11}$$

since $v_1(n)$ and $u(n-m)$ are uncorrelated. If we set $m = 0$ in (4.4.10), it becomes

$$r_u(0) - 0.135 r_u(-1) - 0.7411 r_u(-2) = E\{v_1(n)u(n)\} \tag{4.4.12}$$

If we, next, substitute the value of $u(n)$ from (4.4.8) in (4.4.12), taking into consideration that v and u are independent rv's, we obtain

$$\sigma_1^2 = r_u(0) - 0.135 r_u(1) - 0.7411 r_u(2) \tag{4.4.13}$$

where we used the symmetry property of the correlation function. From the first equation of (4.4.11) we obtain the relation

$$r_u(1) = \frac{0.135}{1-0.7411} r_u(0) = \frac{0.135}{0.2589} \sigma_u^2 \tag{4.4.14}$$

Substituting the above equation in the second equation of the set of (4.4.11), we obtain

$$r_u(2) = 0.135 \frac{0.135}{0.2589} \sigma_u^2 + 0.7411 \sigma_u^2 \tag{4.4.15}$$

Hence, the last three equations give the variance of u

$$\sigma_u^2 = \frac{\sigma_1^2}{0.3282} = \frac{0.31}{0.3282} = 0.9445 \tag{4.4.16}$$

Using (4.4.16), (4.4.14), and the value $\sigma_2^2 = 0.12$, we obtain the correlation matrix

$$\mathbf{R_x} = \mathbf{R_u} + \mathbf{R_{v2}} \begin{bmatrix} 0.9445 & 0.4925 \\ 0.4925 & 0.9445 \end{bmatrix} + \begin{bmatrix} 0.12 & 0 \\ 0 & 0.12 \end{bmatrix} = \begin{bmatrix} 1.0645 & 0.4925 \\ 0.4925 & 1.0645 \end{bmatrix} \tag{4.4.17}$$

From Figure 4.4.2 we find the relation

$$u(n) - 0.931 u(n-1) = d(n) \tag{4.4.18}$$

Multiplying (4.4.18) by $u(n)$ and then by $u(n-1)$, and taking the ensemble averages of the results, we obtain the vector $\mathbf{p}_{dx}$ equals to

$$\mathbf{p}_{dx} = [0.4860 - 0.3868]^T \tag{4.4.19}$$

Minimum mean-square error (MSE)

Introducing the above results in (4.2.3), we obtain the MSE-surface (cost function) as a function of the filter coefficients. Hence,

$$J(\mathbf{w}) = 0.8461 - 2[w_0 \quad w_1]\begin{bmatrix} 0.4860 \\ -0.3868 \end{bmatrix} + [w_0 \quad w_1]\begin{bmatrix} 1.0645 & 0.4925 \\ 0.4925 & 1.0645 \end{bmatrix}$$

$$= 0.8461 + 0.972w_0 - 0.7736w_1 + 1.0645w_0^2 + 1.0645w_1^2 + 0.985w_0w_1 \tag{4.4.20}$$

The MSE surface and its contour plots are shown in Figure 4.4.3.

Optimum filter (w^o)

The optimum filter is defined by (4.3.5), and in this case takes the following form:

$$\mathbf{w}^0 = \mathbf{R}_X^{-1}\mathbf{p}_{dx}\begin{bmatrix} 1.1953 & -0.5531 \\ -0.5531 & 1.1953 \end{bmatrix}\begin{bmatrix} 0.4860 \\ -0.3868 \end{bmatrix} = \begin{bmatrix} 0.7948 \\ -0.7311 \end{bmatrix} = \begin{bmatrix} w_0 \\ w_1 \end{bmatrix} \tag{4.4.21}$$

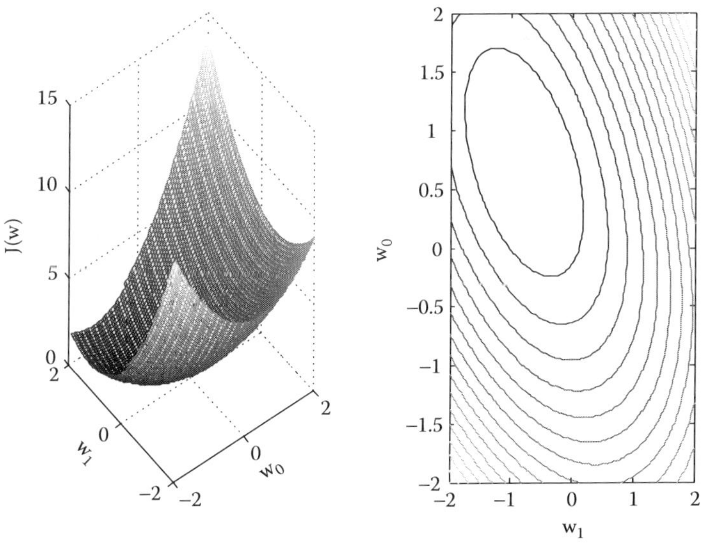

Figure 4.4.3 The MSE surface and its corresponding contour plots of Example 4.4.2.

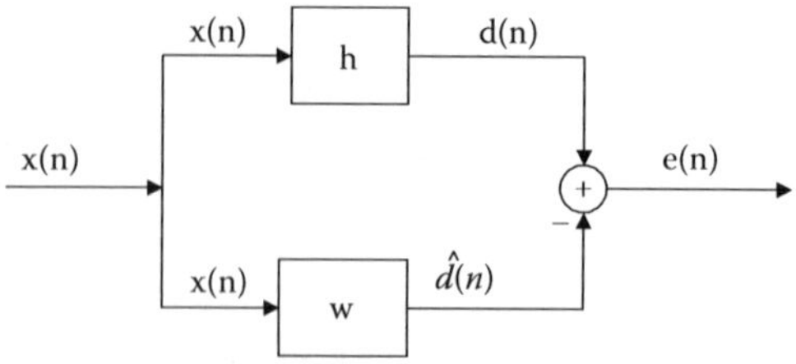

Figure 4.4.4 System identification set up.

The minimum MSE is found using (4.3.6) that, in this case, gives the value

$$J_{\min} = \sigma_d^2 - \mathbf{p}_{\mathbf{dx}}^{\mathrm{T}} \mathbf{R}_{\mathbf{x}}^{-1} \mathbf{p}_{\mathbf{dx}}$$

$$= 0.8461 - [0.4860 \quad -0.3868] \begin{bmatrix} 1.1953 & -0.5531 \\ -0.5531 & 1.1953 \end{bmatrix} \begin{bmatrix} 0.4860 \\ -0.3868 \end{bmatrix}$$

$$= 0.1770 \tag{4.4.22}$$

Example 4.4.3 (System identification): It is desired, using a Wiener filter, to estimate the unknown impulse response coefficients h_i's of a FIR system (see Figure 4.4.4). The input $\{x(n)\}$ is a zero mean iid rv's with variance σ_x^2. Let the impulse response **h** of the filter be: $\mathbf{h} = [0.9 \;\; 0.6 \;\; 0.2]^{\mathrm{T}}$. Since the input $\{x(n)\}$ is zero mean and iid rv's, the correlation matrix $\mathbf{R}_x$ is a diagonal matrix with elements having values σ_x^2. The desired signal $d(n)$ is the output of the unknown filter, and it is given by (see Section 2.4): $d(n) = 0.9x(n) + 0.6x(n-1) + 0.2x(n-2)$. Therefore, the cross-correlation output is given by:

$$p_{dx}(i) = E\{d(n)x(n-i)\} = E\{[0.9x(n) + 0.6x(n-1) + 0.2x(n-2)]x(n-i)\}$$

$$= 0.9E\{x(n)x(n-i)\} + 0.6E\{x(n-1)x(n-i)\} + 0.2E\{x(n-2)x(n-i)\}$$

$$= 0.9r_x(i) + 0.6r_x(i-1) + 0.2r_x(x-2) \tag{4.4.23}$$

Hence, we obtain ($r_x(m) = 0 \;\; for \; m \neq 0$): $p_{dx}(0) = 0.9\sigma_x^2$, $p_{dx}(1) = 0.6\sigma_x^2$. The optimum filter is

$$\mathbf{w}^{\mathbf{o}} = \mathbf{R}_{\mathbf{x}}^{-1}\mathbf{p}_{\mathbf{dx}} = (\sigma_x^2)^{-1} \begin{bmatrix} 1 & 0 \\ 0 & 1 \end{bmatrix} \begin{bmatrix} 0.9 \\ 0.6 \end{bmatrix} = (\sigma_x^2)^{-1} \begin{bmatrix} 0.9 \\ 0.6 \end{bmatrix}$$, and the MMSE is

(assuming $\sigma_x^2 = 1$)

$$J_{\min} = \sigma_d^2 - [0.9 \;\; 0.6]\begin{bmatrix} 1 & 0 \\ 0 & 1 \end{bmatrix}\begin{bmatrix} 0.9 \\ 0.6 \end{bmatrix}.$$

But,

$$\sigma_d^2 = E\{d(n)d(n)\} = E\{[0.9x(n) + 0.6x(n-1) + 0.2x(n-2)]^2\}$$

$$= 0.81 + 0.36 + 0.04 = 1.21 \text{ and, hence, } J_{\min} = 1.21 - (0.9^2 + 0.6^2) = 0.04.$$

Book MATLAB function for system identification (Wiener filter)

```
function[w,jm]=aawienerfirfilter(x,d,M)
%function[w,jm]=aawienerfirfilter(x,d,M);
%x=data entering both the unknown filter(system) and
%the Wiener filter;
%d=the desired signal=output of the unknown system;
length(d)=length(x);
%M=number of coefficients of the Wiener filter;
%w=Wiener filter coefficients;jm=minimum mean-square
%error;
pdx=xcorr(d,x,'biased');
p=pdx(1,(length(pdx) + 1)/2:((length(pdx)+1)/2)+M-1);
rx=aasamplebiasedautoc(x,M);
R=toeplitz(rx);
w=inv(R)*p';
jm=var(d)-p*w;% var() is a MATLAB function;
```

By setting, for example, the following MATLAB procedure: x = randn(1,256); d = filter([0.9 0.2 – 0.4],1,x); [w,jm] = aawienerfirfilter(x,d,4); we obtain: **w** = [0.9000 0.2000 –0.3999 –0.0004], $J_{\min} = 0.0110$. We note that, if we assume a larger number of filter coefficients than those belonging to the unknown system, the Wiener filter produces close approximate values to those in the unknown system and produces values close to zero for the remaining coefficients.

Example 4.4.4 (Noise canceling): In many practical applications there exists a need to cancel the noise added to a signal. For example, we are using the cell phone inside the car and the noise of the car or radio is added to the message we are trying to transmit. A similar circumstance appears when pilots in planes and helicopters try to communicate, or tank drivers try to do the same. Figure 4.4.5 shows pictorially the noise contamination situations. Observe that the noise added to the signal and the other component entering the Wiener filter emanate from the same source but follow different

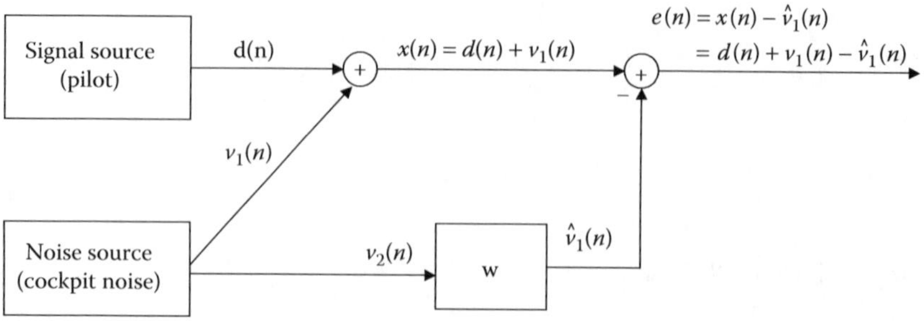

Figure 4.4.5 Illustration of noise canceling scheme.

paths in the same environment. This indicates that there is some degree of correlation between these two noises. It is assumed that the noises have zero mean values. The output of the Wiener filter will approximate the noise added to the desired signal and, thus, the error will be close to the desired signal. The Wiener filter in this case is

$$\mathbf{R}_{\mathbf{v}_2}\mathbf{w}^{o} = \mathbf{p}_{\mathbf{v}_1\mathbf{v}_2} \tag{4.4.24}$$

because the desired signal in this case is $\mathbf{v}_1$.

The individual components of the vector $\mathbf{p}_{\mathbf{v}_1\mathbf{v}_2}$ are

$$\begin{aligned} p_{v_1v_2}(m) &= E\{v_1(n)v_2(n-m)\} = E\{(x(n)-d(n))v_2(n-m)\} \\ &= E\{x(n)v_2(n-m)\} - E\{d(n)v_2(n-m)\} = p_{xv_2}(m) \end{aligned} \tag{4.4.25}$$

Because $d(n)$ and $v_2(n)$ are uncorrelated,

$$E\{d(n)v_2(n-m)\} = E\{d(n)\}E\{v_2(n-m)\} = 0.$$

Therefore, (4.4.24) becomes

$$\mathbf{R}_{\mathbf{v}_2}\mathbf{w}^{o} = \mathbf{p}_{\mathbf{x}\mathbf{v}_2} \tag{4.4.26}$$

To demonstrate the effect of the Wiener filter, let $d(n) = 0.99^n \sin(0.1n\pi + 0.2\pi)$, $v_1(n) = 0.8v_1(n-1) + v(n)$ and $v_2(n) = -0.95v_2(n-1) + v(n)$, where $v(n)$ is white noise with zero mean value and unit variance. The correlation

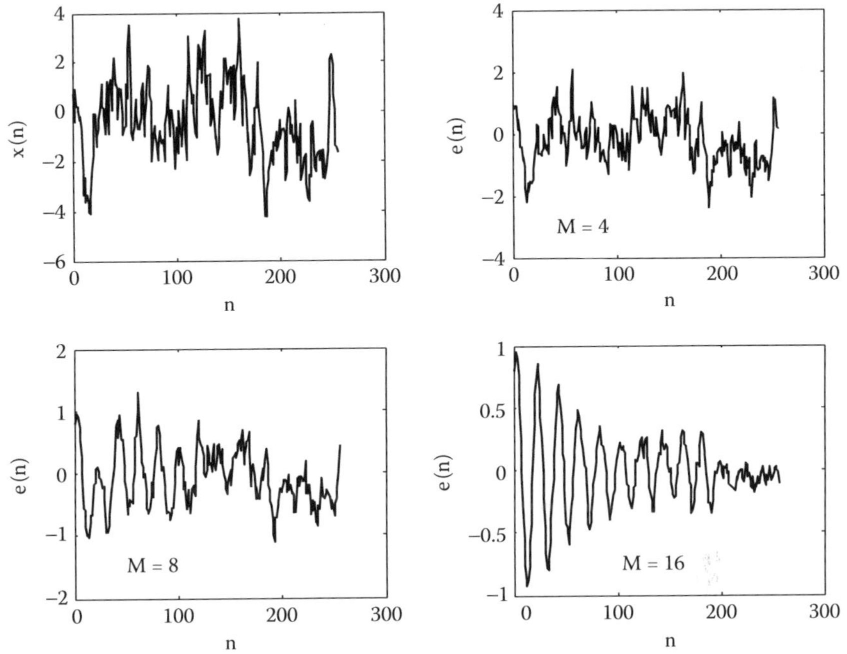

Figure 4.4.6 Illustration of Example 4.4.4.

matrix $\mathbf{R}_{\mathbf{v}_2}$ and cross-correlation vector $\mathbf{p}_{\mathbf{xv}_2}$ are found using the sample biased correlation equations

$$\hat{r}_{v_2v_2}(k) = \frac{1}{N}\sum_{n=0}^{N-1} v_2(n)v_2(n-k) \qquad k = 0, 1, \cdots, K-1, \ K \ll N$$

$$\hat{p}_{xv_2}(k) = \frac{1}{N}\sum_{n=0}^{N-1} x(n)v_2(n-k) \qquad k = 0, 1, \cdots, K-1, \ K \ll N \tag{4.4.27}$$

Figure 4.4.6 shows simulation results for three different-order filters using the Book MATLAB given below.

Book MATLAB function for noise canceling

```
function[d,w,xn]=aawienernoisecancelor(dn,a1,a2,v,M,N)
%[d,w,xn]=aawienernoisecance=
%lor(dn,a1,a2,v,M,N);dn=desired signal;
%a1=first order IIR coefficient,a2=first order IIR
%coefficient;
%v=noise;M=number of Wiener filter coefficients;N=num-
%ber of sequence
```

```
%elemets of dn(desired signal) and v(noise);d=output
%desired signal;
%w=Wiener filter coefficients;xn=corrupted sig-
%nal;en=xn-v1=d;
v1(1)=0;v2(1)=0;
for n=2:N
     v1(n)=a1*v1(n-1)+v(n-1);
     v2(n)=a2*v2(n-1)+v(n-1);
end;
v2autoc=aasamplebiasedautoc(v2,M);
xn=dn +v1;
Rv2=toeplitz(v2autoc);
p1=xcorr(xn,v2,'biased');
if M>N
     disp(['error:M must be less than N']);
end;
R=Rv2(1:M,1:M);
p=p1(1,(length(p1)+1)/2:(length(p1)+1)/2+M-1);
w=inv(R)*p';
yw=filter(w,1,v2);
d=xn-yw(:,1:N);
```

Example 4.4.5 (Self-correcting Wiener filter (SCWF)): We can also arrange the standard single Wiener filter in a series form as shown in Figure 4.4.7. This configuration permits us to process the signal using filters with fewer coefficients, thus saving in computation. Figure 4.4.8a shows the input to the filter, which is a sine wave with added Gaussian white noise. Figure 4.4.8b shows the output of the first stage of a self-correcting Wiener filter and Figure 4.4.8c shows the output of the fourth stage of the self-correcting Wiener filter (each stage has ten coefficients).

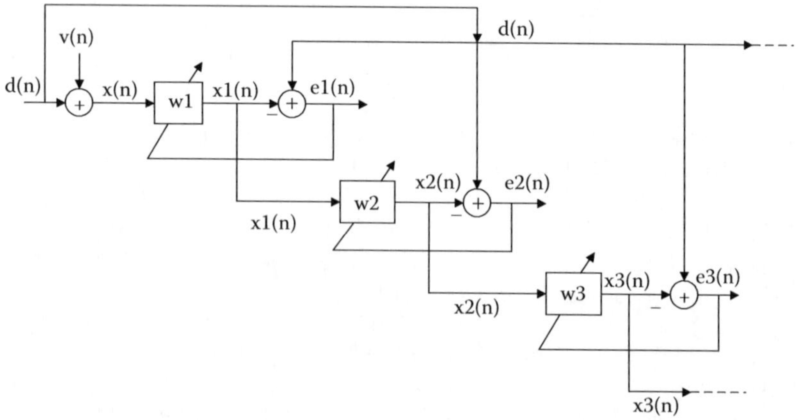

Figure 4.4.7 Self-correcting Wiener filter (SCNF).

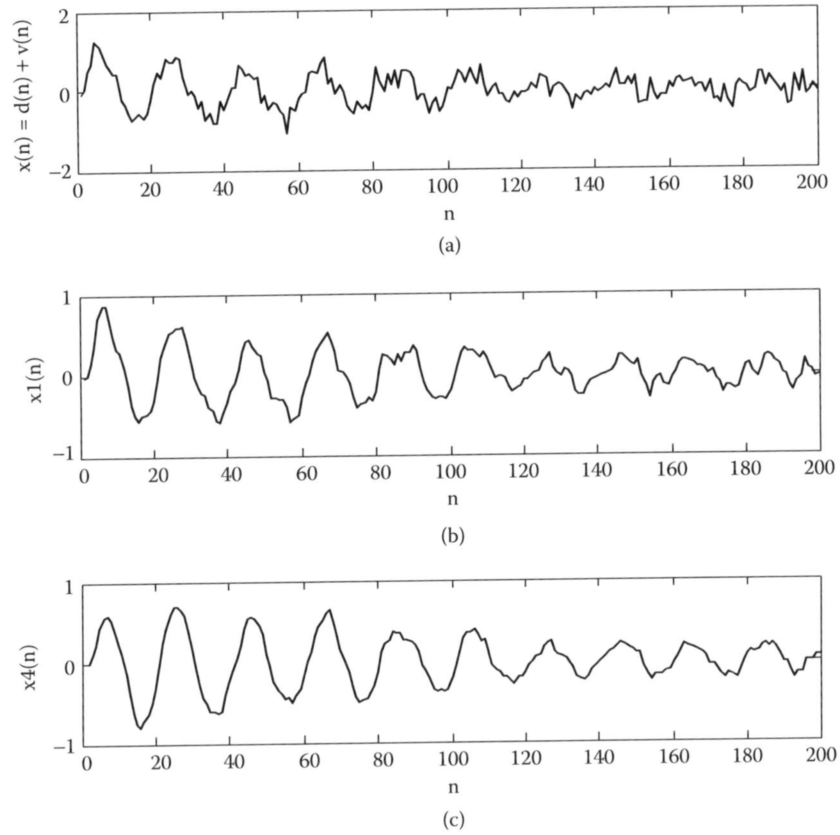

Figure 4.4.8 Illustration of Example 4.4.5.

Problems

4.3.1 Verify (4.3.6).

4.3.2 Find the Wiener coefficients w_0 and w_1 that approximate (models) the unknown system coefficients (see Figure 4.3.1), which are $b_0 = 0.9$ and $b_1 = 0.25$. Let the noise $\{v(n)\}$ be white with zero mean and variance $\sigma_v^2 = 0.15$. Further, we assume that the input data sequence $\{x(n)\}$ is stationary white process with zero mean and variance $\sigma_x^2 = 1$. In addition, $\{v(n)$ and $\{x(n)\}$ are uncorrelated and $\{v(n)\}$ is added to the output of the system under study.

4.3.3 Find $J_{\min}$ using the orthogonality principle.

4.4.1 Let the data entering the Wiener filter are given by $x(n) = d(n) + v(n)$. The noise $v(n)$ has zero mean value, unit variance, and is uncorrelated with the desired signal $d(n)$. Furthermore, assume $r_d(m) = 0.9^m$ and $r_v(m) = \delta(m)$. Find the following: $\mathbf{p}_{dx}$, $\mathbf{w}$, $J_{\min}$, signal power, noise power, signal-to-noise power.

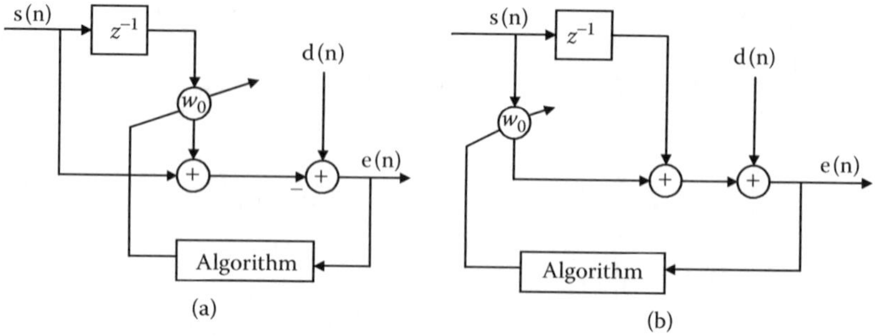

Figure p4.4.3 Wiener filters of Problem 4.4.3.

4.4.2 Repeat Example 4.4.4 with the difference that a small amount of the signal is leaking and is added to $v_2(n)$ noise. Use the following quantities and state your observations: $0.05d(n)$, $0.1d(n)$, $0.3d(n)$, and $0.6d(n)$.

4.4.3 Find the cost function and the MSE surface for the two systems shown in Figure p4.4.3. Given: $E\{s^2(n)\} = 0.9$, $E\{s(n)s(n-1)\} = 0.4$, $E\{d^2(n)\} = 3$, $E\{d(n)s(n)\} = -0.5$, and $E\{d(n)s(n-1)\} = 0.9$.

Hints-solutions-suggestions

4.3.1:

$$J_{\min} = \sigma_d^2 - 2\mathbf{w}^{oT}\mathbf{R}_x\mathbf{w}^o = \sigma_d^2 - \mathbf{w}^{oT}\mathbf{R}_x\mathbf{w}^o = \sigma_d^2 - (\mathbf{R}_x^T\mathbf{w}^o)^T\mathbf{w}^o = \sigma_d^2 - (\mathbf{R}_x\mathbf{w}^o)^T\mathbf{w}^o$$

$$= \sigma_d^2 - \mathbf{p}_{dx}^T\mathbf{w}^o = \sigma_d^2 - \mathbf{p}_{dx}^T\mathbf{R}_x^{-1}\mathbf{p}_{dx} \quad (\mathbf{R}_x \textit{ is symmetric and } \mathbf{w} \textit{ and } \mathbf{p}_{dx} \textit{ are vectors})$$

4.3.2:

$$\mathbf{R}_x = \begin{bmatrix} E\{x^2(n)\} & E\{x(n)x(n-1)\} \\ E\{x(n-1)x(n)\} & E\{x^2(n)\} \end{bmatrix} = \begin{bmatrix} 1 & 0 \\ 0 & 1 \end{bmatrix}$$ (1). Since $\{x(n)\}$ and $\{v(n)\}$ are white processes implies that

$E\{x(n-1)x(n)\} = E\{x(n)x(n-1)\} = E\{x(n-1)x(n)\} = E\{x(n)\}E\{x(n-1)\} = 0$,

$E\{x^2(n)\} = E\{x^2(n-1)\} = 1$, $E\{x(n)v(n)\} = E\{x(n)\}E\{v(n)\} = 0$,

$E\{x(n-1)v(n)\} = E\{x(n-1)\}E\{v(n)\} = 0$,
we obtain

$$\mathbf{p}_{dx} = \begin{bmatrix} E\{[v(n)+0.9x(n)+0.25x(n-1)]x(n)\} \\ E\{[v(n)+0.9x(n)+0.25x(n-1)]x(n-1)\} \end{bmatrix} = \begin{bmatrix} 0.9\sigma_x^2 \\ 0.25\sigma_x^2 \end{bmatrix} = \begin{bmatrix} 0.9 \\ 0.25 \end{bmatrix} \quad (2),$$

$$\sigma_d^2 = E\{d^2(n)\} = E\{[v(n)+0.9x(n)+0.25x(n-1)][v(n)+0.9x(n)+0.25x(n-1)]\}$$

$$= E\{v^2(n)\}+0.81E\{x^2(n)\}+0.0625E\{x^2(n-1)\}=0.15+0.81+0.0625=1.0225$$

(3).

Introducing (1), (2), and (3) in (4.2.3) we obtain

$$J = 1.0225 - 2[w_0 \quad w_1]\begin{bmatrix}0.9\\0.25\end{bmatrix} + [w_0 \quad w_1]\begin{bmatrix}1 & 0\\0 & 1\end{bmatrix}\begin{bmatrix}w_0\\w_1\end{bmatrix}.$$

$$= 1.0225 - 1.8w_0 - 0.5w_1 + w_0^2 + w_1^2$$

The optimum Wiener filter is given by (see (4.3.5))

$$\mathbf{w} = \mathbf{R}_\mathbf{x}^{-1}\mathbf{p}_\mathbf{dx} = \begin{bmatrix}1 & 0\\0 & 1\end{bmatrix}\begin{bmatrix}0.9\\0.25\end{bmatrix} = \begin{bmatrix}0.9\\0.25\end{bmatrix},$$ and the minimum error is given by

$$J_{\min} = \sigma_d^2 - \mathbf{p}_\mathbf{dx}^\mathbf{T}\mathbf{w} = 1.0225-[0.9 \quad 0.25]\begin{bmatrix}0.9\\0.25\end{bmatrix} = 1.0225-0.8725=0.15.$$

4.3.3:

$$J = E\{e(n)[d(n)-\sum_{m=0}^{M-1} w_m x(n-m)]\} = E\{e(n)d(n)\} - \sum_{m=0}^{M-1} w_m E\{e(n)x(n-m)\}.$$ If the coefficients have their optimum value, the orthogonality principle states that $E\{e(n)x(n-m)\}=0$ and,

hence, $J_{\min} = E\{e(n)d(n)\} = E\{d(n)d(n) - \sum_{m=0}^{M-1} w_m^o x(n-m)d(n)\} = \sigma_d^2$

$$-\sum_{m=0}^{M-1} w_m^o E\{d(n)x(n-m)\} = \sigma_d^2 - \sum_{m=0}^{M-1} w_m^o p_{dx}(m) = r_d(0) - \mathbf{p}_\mathbf{dx}^\mathbf{T}\mathbf{w}^\mathbf{o} = r_d(0) - \mathbf{p}_\mathbf{dx}^\mathbf{T}\mathbf{R}_\mathbf{x}^{-1}\mathbf{p}_\mathbf{dx}.$$

4.4.1:

$$p_{dx}(m) = E\{d(n)x(n-m)\} = E\{d(n)d(n-m)\} + E\{(d(n)\}E\{v(n-m)\} = r_d(m) \quad (1)$$

since d(n) and v(n) are independent and WSS. Also, $r_x(m) = E\{x(n)x(n-m)\} = r_d(m)+r_v(m)$, where again the independence and the zero mean properties of the noise were introduced. Therefore, the Wiener equation becomes: $\mathbf{R}_d + \mathbf{R}_v = \mathbf{p}_{dx}$. From the relation $p_{dx}(m) = r_d(m)$ (see (1)), and the given relations in the problem, we find that the Wiener equation and its inverse are given by

$$\begin{bmatrix} 2 & 0.9 \\ 0.9 & 2 \end{bmatrix}\begin{bmatrix} w_0 \\ w_1 \end{bmatrix} = \begin{bmatrix} 1 \\ 0.9 \end{bmatrix}, \quad \begin{bmatrix} w_0 \\ w_1 \end{bmatrix} = \frac{1}{3.19}\begin{bmatrix} 1.19 \\ 0.9 \end{bmatrix} = \begin{bmatrix} 0.373 \\ 0.282 \end{bmatrix}.$$

The minimum error is

$J_{\min} = r_d(0) - \mathbf{p}_{dx}^T \mathbf{R}_x^{-1} \mathbf{w}^o = r_d(0) - \mathbf{p}_{dx}^T \mathbf{w}^o = 0.373$. Since $r_d(0) = \sigma_d^2 = 1$ and $\sigma_v^2 = 1$, the power of the desired signal and noise are equal and, hence, $10\log(1/1) = 0$. After filtering $x(n)$ we find that the signal power is

$E\{\hat{d}^2(0)\} = \mathbf{w}^{oT}\mathbf{R}_d\mathbf{w}^o = [w_0^o \quad w_1^o]\begin{bmatrix} 1 & 0.9 \\ 0.9 & 1 \end{bmatrix}\begin{bmatrix} w_0^o \\ w_1^o \end{bmatrix} = 0.408$, and the noise power is

$$E\{\hat{v}^2(0)\} = \mathbf{w}^{oT}\mathbf{R}_v\mathbf{w}^o = [w_0^o \quad w_1^o]\begin{bmatrix} 1 & 0 \\ 0 & 1 \end{bmatrix}\begin{bmatrix} w_0^o \\ w_1^o \end{bmatrix} = 0.285.$$

Therefore, SNR = 10log(0.408/0.285) = 1.56, which shows that the Wiener filter increased the SNR by 1.56 dB.

4.4.3:

$$\begin{aligned} a) j &= E\{(d(n) - [s(n) + w_0 s(n-1)]^2\} \\ &= E\{d^2(n) - [s(n) + w_0 s(n-1)]^2 - d(n)[s(n) + w_0 s(n-1)]\} \\ &= E\{d^2(n)\} - E\{s^2(n)\} - w_0^2 E\{s^2(n-1)\} - 2w_0 E\{s(n)s(n-1)\} - 2E\{d(n)s(n)\} \\ &\quad - 2w_0 E\{s(n)s(n-1)\} \\ &= 3 - 0.9 - w_0^2 0.9 - 2w_0 0.4 + 2 \times 0.5 - 2w_0 0.4 = 3.1 - 0.9w_0^2 - 0.6w_0 \Rightarrow \end{aligned}$$

$\frac{\partial J}{\partial w_0} = -\,0.9 \times 2w_0 - 1.6 = 0 \Rightarrow w_0 = 0.889$. (b) Similar to (a)

chapter 5

Eigenvalues of R_x — properties of the error surface

5.1 The eigenvalues of the correlation matrix

Let $\mathbf{R}_x$ be an $M \times M$ correlation matrix of a WSS discrete-time process computed from a data vector $\mathbf{x}(n)$. This symmetric and non-negative matrix (see Problem 5.1.1) can be found using MATLAB as follows: we find the data vector $\mathbf{x}(n) = [\,-0.4326 \quad -1.6656 \quad 0.1253 \quad 0.2877 \quad -1.1465 \quad 1.1909 \quad 1.1892 \quad -0.0376 \quad 0.3273 \quad 0.1746\,]$ using the MATLAB command: **x = randn(1, 10).** From this data vector we find its autocorrelation function $r_x(m) = [0.7346 \quad 0.0269 \quad -0.1360]$ using the Book MATLAB function: **rx = aasamplebiasedautoc(x, 3)**. Next, we obtain the correlation matrix

$$\mathbf{R}_x = \begin{bmatrix} 0.7346 & 0.0269 & -0.1360 \\ 0.0269 & 0.7346 & 0.0269 \\ -0.1360 & 0.0269 & 0.7346 \end{bmatrix}$$

using the MATLAB function: $\mathbf{R}_x$ = **toeplitz(rx)**. Having found the correlation matrix $\mathbf{R}_x$, we wish to find an $M \times 1$ vector $\mathbf{q}$ such that

$$\mathbf{R}_x \mathbf{q} = \lambda \mathbf{q} \tag{5.1.1}$$

for some constant λ. The above equation indicates that the left-hand side transformation does not change the direction of the vector $\mathbf{q}$ but only its length. The *characteristic equation* of $\mathbf{R}_x$ is given by

$$\det(\mathbf{R}_x \mathbf{q} - \lambda \mathbf{I}) = 0 \tag{5.1.2}$$

The roots $\lambda_1, \lambda_2, \cdots, \lambda_M$ of the above equation are called *eigenvalues* of $\mathbf{R}_x$. Therefore, each eigenvalue corresponds to an *eigenvector* $\mathbf{q}_i$ such that

$$\mathbf{R}_x \mathbf{q}_i = \lambda_i \mathbf{q}_i \quad i = 1, 2, \cdots, M \tag{5.1.3}$$

The MATLAB function

$$[\mathbf{Q}, \mathbf{D}] = eig(\mathbf{R}); \tag{5.1.4}$$

gives the diagonal matrix $\mathbf{D}$ ($\mathbf{D} = \Lambda$) containing the eigenvalues of $\mathbf{R}$ ($\mathbf{R} = \mathbf{R}_x$) and gives the matrix $\mathbf{Q}$ with its columns equal to the eigenvectors of $\mathbf{R}$. Using the 3 × 3 matrix above in (5.1.4), we obtain

$$\mathbf{Q} = \begin{bmatrix} 0.6842 & 0.1783 & -0.7071 \\ -0.2522 & 0.9677 & -0.0000 \\ 0.6842 & 0.1783 & 0.7071 \end{bmatrix}, \quad \mathbf{D} = \begin{bmatrix} 0.5886 & 0 & 0 \\ 0 & 0.7445 & 0 \\ 0 & 0 & 0.8706 \end{bmatrix}$$

In Table 5.1.1 below, we give the eigenvalue properties of WSS processes. Continuing with our correlation matrix using MATLAB, we also find that

```
>>Q(:,1)'*Q(:,3) = 5.5511e - 17 % Q (:, 1)'= row of the first column
                                % of Q, Q (1:,3) = third column of Q,
                                % the results verify the orthogonality
                                % property;
```

Table 5.1.1 Eigenvalue Properties

$\mathbf{x}(n) = [x(n)\ x(n-1) \cdots x(n-M+1)]$	Wide-sense stationary stochastic process
$\mathbf{R}_x = E\{\mathbf{x}(n)\mathbf{x}^T(n)\}$	Correlation matrix
λ_i	The eigenvalues of $\mathbf{R}_x$ are real and positive
$\mathbf{q}_i^T \mathbf{q}_j = 0$	Two eigenvectors belonging to two different eigenvalues are orthogonal
$\mathbf{Q}^T\mathbf{Q} = \mathbf{I}$	$\mathbf{Q} = [\mathbf{q}_0\ \mathbf{q}_1 \cdots \mathbf{q}_{M-1}]$; $\mathbf{Q}$ is a unitary matrix
$\mathbf{R}_x = \mathbf{Q}\Lambda\mathbf{Q}^T = \sum_{i=0}^{M-1} \lambda_i \mathbf{q}_i \mathbf{q}_i^T$	$\Lambda = diag[\lambda_0\ \lambda_1\ \lambda_2 \cdots \lambda_{M-1}]$
$tr\{\mathbf{R}_x\} = \sum_{i=0}^{M-1} \lambda_i$	$tr\{\mathbf{R}\}$ = trace of $\mathbf{R}$ = sum of the diagonal elements of $\mathbf{R}$
$\lambda_{max} \le S_x^{max} = \max_{-\pi \le \omega \le \pi} S_x(e^{j\omega})$	
$\lambda_{min} \ge S_x^{min} = \min_{-\pi \le \omega \le \pi} S_x(e^{j\omega})$	

$$\mathbf{Q} * \Lambda * \mathbf{Q}^T = \begin{bmatrix} 0.7346 & 0.0269 & -0.1360 \\ 0.0269 & 0.7346 & 0.0269 \\ -0.1360 & 0.0269 & 0.7346 \end{bmatrix} = \mathbf{R}_x \text{ as it should be.}$$

5.2 Geometrical properties of the error surface

The cost function J (see Section 4.2) can be written in the form

$$\mathbf{w}^T \mathbf{R}_x \mathbf{w} - 2\mathbf{p}^T \mathbf{w} - (J - \sigma_d^2) = 0 \qquad (5.2.1)$$

If we set values of $J > J_{min}$, the **w** plane will cut the second order surface, for a two-coefficient filter, along a line whose projection on the **w**-plane are ellipses arbitrarily oriented as shown in Figure 5.2.1. To obtain the contours, we used the following Book MATLAB program:

```
w0=-3:0.05:3;
w1=-3:0.05:3;
[x,y]=meshgrid(w0,w1);
j=0.8461+0.972*x-0.773*y+1.0647*x.^2+1.064*y.^2+0.985x.*y;
contour(x,y,j,30);%30 is the number of desired contours
```

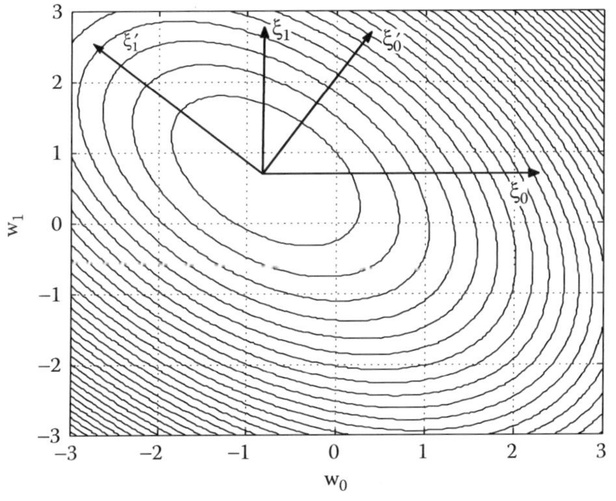

Figure 5.2.1 MSE contours in the **w**- and ξ-planes.

If we introduce the MATLAB function **contour(j, [2.3 3.1 5])** in the command window, we produce three contours at heights 2.3, 3.1, and 5.

The first simplification we can do is to shift the origin of the **w**-axes to another one whose origin is on the center of the ellipses. This is accomplished using the transformation: $\xi = \mathbf{w} - \mathbf{w}^o$. If we introduce this transformation in (5.2.1) and setting $J = 2J_{\min}$, we obtain the relationship (see Problem 5.2.1)

$$\xi^{\mathrm{T}}\mathbf{R}_{\mathbf{x}}\xi - 2J_{\min} + \sigma_d^2 - \mathbf{p}^{\mathrm{T}}\mathbf{w}^o = 0 \tag{5.2.2}$$

But $\sigma_d^2 - \mathbf{p}^{\mathrm{T}}\mathbf{w}^o = J_{\min}$, and thus, the above equation becomes

$$\xi^{\mathrm{T}}\mathbf{R}_{\mathbf{x}}\xi = J_{\min} \tag{5.2.3}$$

We can further simplify (5.2.3) by a linear transformation such that the major axes are aligned with a new set of axes, (ξ_0', ξ_1'). The transformation is accomplished by the linear relationship

$$\xi = \mathbf{Q}\xi' \quad \text{or} \quad \xi' = \mathbf{Q}^{\mathrm{T}}\xi \tag{5.2.4}$$

where **Q** is a matrix whose columns are the eigenvectors of the correlation matrix $\mathbf{R}_\mathbf{x}$. The matrix **Q** is *unitary* having the property: $\mathbf{Q}^{\mathrm{T}} = \mathbf{Q}^{-1}$ (see Table 5.1.1). Substituting next (5.2.4) in (5.2.3) and setting $\mathbf{R}_{\mathbf{x}} = \mathbf{Q}\Lambda\mathbf{Q}^{\mathrm{T}}$ or $\mathbf{Q}^{\mathrm{T}}\mathbf{R}_{\mathbf{x}}\mathbf{Q} = \Lambda$, we obtain the equation

$$\xi'^{\mathrm{T}}\Lambda\xi' = J_{\min} \tag{5.2.5}$$

The matrix is a diagonal matrix whose elements are the eigenvalues of the correlation matrix $\mathbf{R}_\mathbf{x}$.

Example 5.2.1: Let $\lambda_1 = 1$, $\lambda_2 = 0.5$, and $J_{\min} = 0.67$. The ellipse in the (ξ_0', ξ_0') plane is found by solving the system

$$[\xi_0' \quad \xi_0'] \begin{bmatrix} 1 & 0 \\ 0 & 0.5 \end{bmatrix} \begin{bmatrix} \xi_0' \\ \xi_1' \end{bmatrix} = 0.67 \quad or \quad \left(\frac{\xi_0'}{\sqrt{0.67/1}} \right)^2 + \left(\frac{\xi_1'}{\sqrt{0.67/0.5}} \right)^2 = 1 \tag{5.2.6}$$

where 0.67/0.5 is the major axis and 0.67/1 is the minor one. Hence, for the case $J = 2J_{\min}$, the ellipse intersects the ξ_0' axis at 0.67 and the ξ_1' at 1.34.

If we start with (5.2.1) and apply the shift and rotation transformations, we obtain the relationships

$$\begin{aligned} J &= J_{\min} + (\mathbf{w} - \mathbf{w}^o)^T \mathbf{R}_x (\mathbf{w} - \mathbf{w}^o) = J_{\min} + \xi^T \mathbf{R} \xi \\ &= J_{\min} + \xi^T (\mathbf{Q} \Lambda \mathbf{Q}^T) \xi = J_{\min} + \xi'^T \Lambda \xi' \end{aligned} \tag{5.2.7}$$

Notes:

- The contours intersect the ξ'-axes at values dependent upon the eigenvalues of $\mathbf{R}_x$ and the specific MSE value chosen. The rotation and translation do not alter the shape of the MSE surface.
- If the successive contours for the values $2J_{\min}$, $3J_{\min}$, etc., are close to each other, the surface is steep, which in turn indicates that the mean-square estimation error is very sensitive to the choice of the filter coefficients.
- Choosing the filter values $\mathbf{w}$ is equivalent to choosing a point in the $\mathbf{w}$-plane. The height of the MSE surface above the plane at that point is determined only by the signal correlation properties.

Problems

5.1.1 Prove that a correlation matrix from a WSS process is positive definite. That is, $\mathbf{a}^T\mathbf{R}\mathbf{a} \geq 0$ for any vector $\mathbf{a}$.

5.1.2 Show that if $\lambda_1, \lambda_2, \ldots \lambda_M$ denote the eigenvalues of the correlation matrix $\mathbf{R}_x$, then the eigenvalues of $\mathbf{R}^k$ are equal to $\lambda_1^k, \lambda_2^k, \cdots, \lambda_M^k$ for any $k > 0$.

5.1.3 Show that if the eigenvectors $\mathbf{q}_1, \mathbf{q}_2, \ldots \mathbf{q}_M$ correspond to distinct eigenvalues $\lambda_1, \lambda_2, \ldots, \lambda_M$ of an $M \times M$ matrix $\mathbf{R}_x$ then the eigenvectors are linearly independent.

5.1.4 Show that if the $M \times M$ correlation matrix $\mathbf{R}_x$ has M eigenvalues, these eigenvalues are real and non-negative.

5.1.5 Show that if the correlation matrix $\mathbf{R}_x$ has M distinct eigenvalues and eigenvectors then the eigenvectors are orthogonal to each other.

5.1.6 Show that an $M \times M$ correlation matrix $\mathbf{R}_x$ can be written in the form $\mathbf{Q}^H\mathbf{R}_x\mathbf{Q} = \Lambda$, where $\mathbf{Q} = [\mathbf{q}_1 \ \mathbf{q}_2 \ \ldots \ \mathbf{q}_M]$ is a matrix whose columns are the eigenvectors of the correlation matrix and Λ is a diagonal matrix whose elements are the distinct eigenvalues of the correlation matrix.

5.1.7 Show that the trace of the correlation matrix is equal to the sum of its eigenvalues.

5.2.1 Verify (5.2.2).

5.2.2 Find the ellipses if the following data are given:

$$\mathbf{R}_x = \begin{bmatrix} 2 & 1 \\ 1 & 3 \end{bmatrix}, \mathbf{p}_{dx} = \begin{bmatrix} 6 \\ 7 \end{bmatrix}, \sigma_d^2 = 28.$$

5.2.3 A Wiener filter is characterized by the following parameters: $\mathbf{R} = \begin{bmatrix} d & a \\ a & d \end{bmatrix}, \mathbf{p} = \begin{bmatrix} 1 \\ 1 \end{bmatrix}$. $\sigma_d^2 = 2$. It is requested to explore the performance surface as the ratio λ_0/λ_1 varies. This is accomplished using different values of a and b.

Hints-solutions-suggestions

5.1.1:

$a^T\mathbf{R}a \geq 0 = a^TE\{xx^T\}a = E\{(a^Tx)(x^Ta)\} = E\{(a^Tx)^2\} \geq 0$

5.1.2:

Repeated pre-multiplication of both sides of (5.1.1) by the matrix $\mathbf{R}_x$ yields $\mathbf{R}_x{}^k\mathbf{q} = \lambda^k\mathbf{q}$. This shows that if λ is an eigenvalue of $\mathbf{R}_x$, then λ^k is an eigenvalue of $\mathbf{R}^k$ and every eigenvector of $\mathbf{R}_x$ is an eigenvector of $\mathbf{R}^k$.

5.1.3:

Suppose that $\sum_{i=1}^{M} \alpha_i \mathbf{q}_i = \mathbf{0}$ holds for certain not all zero scalars a_i's. Repeated multiplication of this equation and using results of Problem 5.1.2, the following set of M equations are found: $\sum_{i=1}^{M} \alpha_i \lambda_i^{k-1} \mathbf{q}_1 = \mathbf{0}, k = 1, 2, \cdots, M$. This set of equations may be written as follows: $[\alpha_1\mathbf{q}_1 \ \ \alpha_2\mathbf{q}_2 \ \cdots \ \alpha_M\mathbf{q}_M]\mathbf{V} = \mathbf{0}$ (1).

The matrix $\mathbf{V} = \begin{bmatrix} 1 & \lambda_1 & \lambda_1^2 & \cdots & \lambda_1^{M-1} \\ 1 & \lambda_2 & \lambda_2^2 & \cdots & \lambda_2^{M-1} \\ & & \vdots & & \\ 1 & \lambda_M & \lambda_M^2 & \cdots & \lambda_M^{M-1} \end{bmatrix}$ is a *Vandermonde* matrix, which is nonsingular (has an inverse) if the eigenvalues are distinct. Next, we post-multiply (1) by $\mathbf{V}^{-1}$ to obtain $[\alpha_1\mathbf{q}_1 \ \ \alpha_2\mathbf{q}_2 \ \cdots \ \alpha_M\mathbf{q}_M] = \mathbf{0}$. Equating the corresponding columns, we find that $\alpha_i\mathbf{q}_i = \mathbf{0}$. Since q_i's are nonzero, the equations can be satisfied if the a_i's are zero. This contradicts the assumption, and thus, the eigenvectors are linearly independent.

5.1.4:

The ith eigenvalue is found from the relation $\mathbf{R}\mathbf{q}_i = \lambda_i\mathbf{q}_i$ for $i = 1, 2, \ldots M(1)$. Pre-multiplying (1) by $\mathbf{q}_i^H$ (H stands for the Hermitian transpose and is substituted with T for real symmetric matrices) we find $\mathbf{q}_i^H\mathbf{R}_x\mathbf{q}_i = \lambda_i\mathbf{q}_i^H\mathbf{q}_i$.

Because the matrix is positive definite, the left-hand side of the equation is positive or equal to zero. The eigenvalue product on the right-hand side is the Euclidean norm (distance), and it is positive. Hence, the eigenvalues are positive or zero (non-negative).

5.1.5:
Two eigenvectors are orthogonal if $\mathbf{q}_i^H \mathbf{q}_j = 0 \; i \neq j$. From the eigenvalue-eigenvector relationship we write: $\mathbf{R}_x \mathbf{q}_i = \lambda_i \mathbf{q}_i$ (1) $\mathbf{R}_x \mathbf{q}_j = \lambda_j \mathbf{q}_j$ (2). Multiplying (1) by the Hermitian form of the *j* eigenvector, we obtain $\mathbf{q}_j^H \mathbf{R}_x \mathbf{q}_i = \lambda_i \mathbf{q}_j^H \mathbf{q}_i$ (3). Next we take the conjugate transpose of (2) and incorporate the conjugate transpose equivalence property of the correlation matrix. We, next, multiply these results with $\mathbf{q}_i$ from the right to find the relation $\mathbf{q}_j^H \mathbf{R}_x \mathbf{q}_i = \lambda_i \mathbf{q}_j^H \mathbf{q}_i$ (4). Subtracting (3) from (4) we obtain the relation $(\lambda_i - \lambda_j)\mathbf{q}_j^H \mathbf{q}_i = 0$ (5). Because the eigenvalues are distinct, (5) implies that the vectors are orthogonal.

5.1.6:
We assume that the eigenvectors are orthonormal (see Problem 5.1.5). For each eigenvalue and the corresponding eigenvector, there is the relationship $\mathbf{R}_x \mathbf{q}_i = \lambda_i \mathbf{q}_i, i = 1, 2, \cdots, M$ (1). Hence, the set of equations of (1) can be written in the form $\mathbf{R}_x \mathbf{Q} = \mathbf{Q}\Lambda$ (2) because the eigenvectors are orthonormal we find that $\mathbf{Q}^H\mathbf{Q} = \mathbf{I}$ (3). Equivalently we may write (3) as follows: $\mathbf{Q}^{-1} = \mathbf{Q}^H$ (4). If a matrix obeys condition (4) we say that the matrix is *unitary.* Multiplying (2) by $\mathbf{Q}^H$ from the left, we obtain the relation: $\mathbf{Q}^H\mathbf{R}\mathbf{Q} = \Lambda$. This transformation is called the *unitary similarity transformation*.

5.1.7:

$$tr\{\mathbf{Q^H R_x Q}\} = tr\{\Lambda\} = \sum_{i=1}^{M} \lambda_i = tr\{\mathbf{R_x Q^H Q}\} = tr\{\mathbf{R_x I}\} = tr\{\mathbf{R}\}$$

5.2.1:
$(\mathbf{w} - \mathbf{w}^o + \mathbf{w}^o)^T\mathbf{R}(\mathbf{w} - \mathbf{w}^o + \mathbf{w}^o) - 2\mathbf{p}^T(\mathbf{w} - \mathbf{w}^o + \mathbf{w}^o) - 2J_{min} + \sigma_d^2 = (\xi+\mathbf{w}^o)^T \mathrm{R}(\xi+\mathbf{w}^o) - 2\mathbf{p}^T(\xi+\mathbf{w}^o) - 2J_{min} + \sigma_d^2$. With $\mathbf{R}\mathbf{w}^o = \mathbf{p}$, $\mathbf{w}^{oT}\mathbf{p} = \mathbf{p}^T\mathbf{w}^o$ = number and $\sigma_d^2 - \mathbf{p}^T\mathbf{w}^o = J_{min}$ the equation is easily proved.

5.2.2:

$$J = J_{min} + \xi'^T\Lambda\xi' \text{ and } J_{min} = \sigma_d^2 - \mathbf{p}^T\mathbf{w}^o, \; |\mathbf{R} - \lambda\mathbf{I}| = \begin{vmatrix} 2-\lambda & 1 \\ 1 & 3-\lambda \end{vmatrix} = (2-\lambda)(3-\lambda) - 1 = 0,$$

$\lambda_1 = 1.382$,

$$\lambda_2 = 3.618 \begin{bmatrix} 2 & 1 \\ 1 & 3 \end{bmatrix} \begin{bmatrix} w_0^o \\ w_1^o \end{bmatrix} = \begin{bmatrix} 6 \\ 7 \end{bmatrix}, \; \begin{bmatrix} w_0^o \\ w_1^o \end{bmatrix} = \begin{bmatrix} 11/5 \\ 8/5 \end{bmatrix}, J_{min} = 28 - [6 \;\; 7] \begin{bmatrix} 11/5 \\ 8/5 \end{bmatrix} = 3.6$$

$$J = 3.6 + [\xi_0' \;\; \xi_1'] \begin{bmatrix} 1.328 & 0 \\ 0 & 3.618 \end{bmatrix} \begin{bmatrix} \xi_0^o \\ \xi_1^o \end{bmatrix} = 3.6 + 1.382\xi_0' + 3.618\xi_1'$$

5.2.3:

The MSE surface is given by: $J = 2 - 2[w_0 \; w_1]\begin{bmatrix}1\\1\end{bmatrix} + [w_0 \; w_1]\begin{bmatrix}d & a\\a & d\end{bmatrix}\begin{bmatrix}w_0\\w_1\end{bmatrix}$, and the optimum tap weights are: $\begin{bmatrix}w_0^o\\w_1^o\end{bmatrix} = \mathbf{R}^{-1}\mathbf{p} = \begin{bmatrix}d & a\\a & d\end{bmatrix}^{-1}\begin{bmatrix}1\\1\end{bmatrix} = \begin{bmatrix}\frac{1}{d+a}\\ \frac{1}{d+a}\end{bmatrix}$. Therefore, the minimum MSE surface is: $J_{\min} = \sigma_d^2 - \mathbf{w}^{oT}\mathbf{p} = 2 - [\frac{1}{d+a} \;\; \frac{1}{d+a}]\begin{bmatrix}1\\1\end{bmatrix} = 2\frac{d+a-1}{d+a}$.

We also have the relation

$$J = J_{\min} + (\mathbf{w} - \mathbf{w}^o)^T\mathbf{R}(\mathbf{w} - \mathbf{w}^o) = 2\frac{d+a+1}{d+a} + [\xi_0 \;\; \xi_1]\begin{bmatrix}d & a\\a & d\end{bmatrix}\begin{bmatrix}\xi_0\\ \xi_1\end{bmatrix}.$$

From the relation $\begin{vmatrix}d-\lambda & a\\a & d-\lambda\end{vmatrix} = 0$, we obtain the two eigenvalues $\lambda_0 = d + a$, and $\lambda_1 = d - a$. By incorporating different values for a and d we can create different ratios of the eigenvalues. The larger the ratio the more elongated the ellipses are.

chapter 6

Newton and steepest-descent method

6.1 One-dimensional gradient search method

In general, we can say that adaptive algorithms are nothing but iterative search algorithms derived from minimizing a cost function with the true statistics replaced by their estimates. To study the adaptive algorithms, it is necessary to have a thorough understanding of the iterative algorithms and their convergence properties. In this chapter we discuss the *steepest descent method* and the *Newton's method.*

The one-coefficient MSE surface (line) is given by (see (5.2.7))

$$J(w) = J_{\min} + r_x(0)(w - w^o)^2 \tag{6.1.1}$$

and it is pictorially shown in Figure 6.1.1. The first and second derivatives are

$$\frac{\partial J(w)}{\partial w} = 2r(0)(w - w^o) \quad a); \quad \frac{\partial^2 J(w)}{\partial w^2} = 2r_x(0) > 0 \quad b) \tag{6.1.2}$$

Since at $w = w^o$ the first derivative is zero and the second derivative is greater than zero, the surface has a global minimum and is concave upward. To find the optimum value of w, we can use an iterative approach. We start with an arbitrary value $w(0)$ and measure the slope of the curve $J(w)$ at $w(0)$. Next, we set $w(1)$ to be equal to $w(0)$ plus the negative of an increment proportional to the slope of $J(w)$ at $w(0)$. Proceeding with the iteration, we will eventually find the minimum value w^o. The values $w(0)$, $w(1)$, … , are known as the gradient estimates.

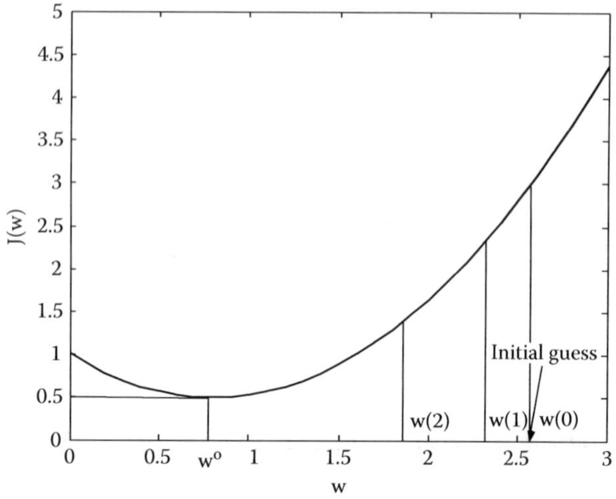

Figure 6.1.1 Gradient search of one-dimensional MSE surface.

Gradient search algorithm

Based on the above development, the filter coefficient at iteration n, $w(n)$, is found using the relation

$$\begin{aligned} w(n+1) &= w(n) + \mu\left(-\frac{\partial J(w)}{\partial w}\bigg|_{w=w(n)}\right) \\ &= w(n) + \mu[-\nabla J(w(n))] = w(n) - 2\mu r_x(0)(w(n) - w^o) \end{aligned} \tag{6.1.3}$$

where μ is a constant to be determined. Rearranging (6.1.3), we find

$$w(n+1) = (1 - 2\mu r_x(0))w(n) + 2\mu r_x(0)w^o \tag{6.1.4}$$

The solution of the above difference equation, using the iteration approach (see Problem 6.1.1), is

$$w(n) = w^o + (1 - 2\mu r_x(0))^n (w(0) - w^o) \tag{6.1.5}$$

The above equation gives $w(n)$ explicitly at any iteration in the search procedure. This is the solution to the *gradient search algorithm*. Note that if we had initially guessed $w(0) = w^o$, which is the optimum value, we would have found $w(1) = w^o$, and this gives the optimum value in one step.

To have convergence of $w(n)$ in (6.1.5) we must impose the condition

$$|1 - 2\mu r_x(0)| < 1 \tag{6.1.6}$$

The above inequality defines the range of the step-size constant μ so that the algorithm will converge. Hence, we obtain

$$-1 < 1 - 2\mu r_x(0) < 1 \text{ or } 0 < 2\mu r_x(0) < 2 \text{ or } 0 < \mu < \frac{1}{r_x(0)} \tag{6.1.7}$$

Under this condition, (6.1.5) converges to the optimum value w^o as $n \to \infty$. If $\mu > 1/r_x(0)$, the process is unstable and no convergence takes place.

When the filter coefficient has a value $w(n)$ (i.e., at iteration n), then the MSE surface (here a line) is (see (6.1.1))

$$J(n) = J_{\min} + r_x(0)(w(n) - w^o)^2 \tag{6.1.8}$$

Substituting the quantity $w(n) - w^o$ of (6.1.5) in (6.1.8) we obtain

$$J(n) = J_{\min} + r_x(0)(w(0) - w^o)^2(1 - 2\mu r_x(0))^{2n} \tag{6.1.9}$$

The above two equations show that $w(n)w^o$ as n increases to infinity, and the MSE undergoes a geometric progression toward $J_{\min}$. The plot of the performance surface $J(n)$ vs. the iteration number n is known as the *learning curve.*

Newton's method in gradient search

Newton's method finds the solution (zeros) to the equation $f(w) = 0$. From Figure 6.1.2 we observe that the slope at $w(0)$ is

$$f'(w(0)) = \frac{df(w)}{dw}\bigg|_{w=w(0)} = \frac{f(w(0))}{w(0) - w(1)} \tag{6.1.10}$$

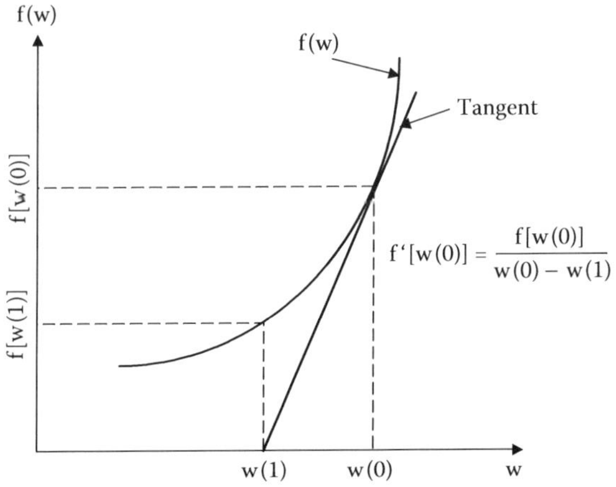

Figure 6.1.2 Newton's method for finding a zero of $f(w)$.

where $w(0)$ is an initial guess. The above equation is the result of retaining the first two terms of Taylor's expansion and setting the rest of them equal to zero. This equation can be written in the form

$$w(1) = w(0) - \frac{f(w(0))}{f'(w(0))} \tag{6.1.11}$$

From (6.1.11), it is clear that if we know the value $w(0)$ and the values of the function and its derivative at the same point, we can find $w(1)$. Hence, the nth iteration of the above equation takes the form

$$w(n+1) = w(n) - \frac{f(w(n))}{f'(w(n))} \qquad n = 0,1,2\cdots \tag{6.1.12}$$

But $f'(w(n)) \cong [f(w(n)) - f(w(n-1))]/[w(n) - w(n-1)]$, and hence, (6.1.12) becomes

$$w(n+1) = w(n) - \frac{f(w(n))[w(n) - w(n-1)]}{f(w(n)) - f(w(n-1))} \tag{6.1.13}$$

As we have mentioned above, Newton's method finds the roots of the function $f(w)$. That is solving the polynomial $f(w) = 0$. However, in our case we need to find the minimum point of the performance surface (here line). This is equivalent to setting $\partial J(w)/\partial w = 0$. Therefore, we substitute the derivative $\partial J(w)/\partial w$ for $f(w)$ in (6.1.12) and for $f'(w)$ we substitute the second order derivative $\partial^2 J(w)/\partial w^2$. Hence, (6.1.12) becomes

$$w(n+1) = w(n) - \frac{\dfrac{\partial J(w(n))}{\partial w(n)}}{\dfrac{\partial^2 (w(n))}{\partial w^2(n)}} \qquad n = 0,1,2\cdots \tag{6.1.14}$$

Newton's multidimensional case

In the previous chapter, it was found that the optimum filter is given by (see (4.3.5))

$$\mathbf{w^o} = \mathbf{R_x^{-1} p_{dx}} \tag{6.1.15}$$

Furthermore, (4.3.1a), (4.3.3a), and (4.3.3b) can be written in the form

$$\begin{bmatrix} \dfrac{\partial J(\mathbf{w})}{\partial w_0} \\ \\ \dfrac{\partial J(\mathbf{w})}{\partial (w_1)} \end{bmatrix} = 2\begin{bmatrix} r_x(0) & r_x(1) \\ r_x(1) & r_x(0) \end{bmatrix}\begin{bmatrix} w_0 \\ w_1 \end{bmatrix} - 2\begin{bmatrix} r_{dx}(0) \\ r_{dx}(1) \end{bmatrix} \tag{6.1.16}$$

or equivalently for M coefficients in the vector form ($\mathbf{r}_{dx} = \mathbf{p}_{dx}$)

$$\nabla J(\mathbf{w}) = 2\mathbf{R}_x\mathbf{w} - 2\mathbf{p}_{dx}, \quad \nabla = \begin{bmatrix} \frac{\partial}{\partial w_0} \\ \frac{\partial}{\partial w_1} \\ \vdots \\ \frac{\partial}{\partial w_M} \end{bmatrix} \tag{6.1.17}$$

The reader should refer to Appendix A for vector and matrix differentiations.

Multiplying (6.1.17) by the inverse of the correlation matrix and using (6.1.15), we find

$$\mathbf{w}^o = \mathbf{w} - \mu\mathbf{R}_x^{-1}\nabla J(\mathbf{w}) \tag{6.1.18}$$

where we have substituted the constant 1/2 with the step-size parameter. If we introduce the new gradient $\mathbf{R}_x^{-1}\nabla J(\mathbf{w})$ in (6.1.3), we obtain

$$\mathbf{w}(n+1) = \mathbf{w}(n) - \mu\mathbf{R}_x^{-1}\nabla J(\mathbf{w}(n)) \tag{6.1.19}$$

This equation is a generalization of (6.1.3). Based on the observed correspondence between quantities we can substitute w's with their vector equivalents, the autocorrelation $r_x(0)$ with the correlation matrix $\mathbf{R}_x$, and $2r_x(0)(w(n) - w^o)$ with the gradient vector of $J(\mathbf{w})$. Hence, we obtain

$$\mathbf{w}(n+1) = \mathbf{w}(n) - 2\mu\mathbf{R}_x(\mathbf{w}(n) - \mathbf{w}^o) = \mathbf{w}(n) - \mu\nabla J(\mathbf{w}) \tag{6.1.20}$$

The presence of $\mathbf{R}_x$ causes the eigenvalue spread problem to be exaggerated. This is because, if the ratio of the maximum to the minimum eigenvalues (i.e., *eigenvalue spread*) is large, the inverse of the correlation matrix may have large elements, and this will result in difficulties in solving equations that involve inverse correlation matrices. In such cases, we say that the correlation matrices are *ill-conditioned*.

To overcome the eigenvalue spread problem, Newton's method simply substitutes μ with $\mu\mathbf{R}_x^{-1}$ in (6.1.20) to obtain (6.1.19). This substitution has the effect of rotating the gradient vector to the direction pointing toward the minimum point of the MSE surface as shown in Figure 6.1.3.

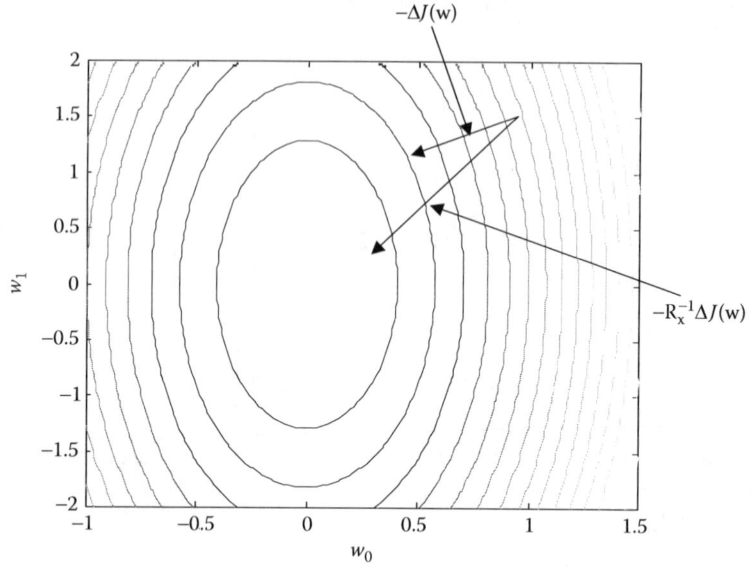

Figure 6.1.3 The gradient and Newton vectors.

Substituting (6.1.17) in (6.1.19) we obtain

$$\begin{aligned}\mathbf{w}(n+1) &= \mathbf{w}(n) - \mu\mathbf{R}_x^{-1}(2\mathbf{R}_x\mathbf{w}(n) - 2\mathbf{p}_{dx}) = (1-2\mu)\mathbf{w}(n) + 2\mathbf{R}_x^{-1}\mathbf{p}_{dx} \\ &= (1-2\mu)\mathbf{w}(n) + 2\mu\mathbf{w}^o\end{aligned} \quad (6.1.21)$$

Subtracting $\mathbf{w}^o$ from both sides of the above equation we find

$$\mathbf{w}(n+1) - \mathbf{w}^o = (1-2\mu)(\mathbf{w}(n) - \mathbf{w}^o) \quad (6.1.22)$$

For $n = 0$ and $\mu = 1/2$, we obtain $\mathbf{w}^o$ in one step. However, in practice $\nabla J(w)$ and $\mathbf{R}_x^{-1}$ are estimated and, therefore, the value of the step-size parameter must be less than 0.5.

Setting $\mathbf{w} - \mathbf{w}^o = \xi$ in (6.1.22) we find the relation

$$\xi(n+1) = (1-2\mu)\xi(n) \quad (6.1.23)$$

which has the solution (see also Problem 6.1.1)

$$\xi(n) = (1-2\mu)^n\xi(0) \quad or \quad \mathbf{w}(n) - \mathbf{w}^o = (1-2\mu)^n(\mathbf{w}(0) - \mathbf{w}^o) \quad (6.1.24)$$

Using (5.2.7) in connection with (6.1.24) we obtain (see Problem 6.1.3)

$$J(n) = J_{min} + (1-2\mu)^{2n}(J(0) - J_{min}) \tag{6.1.25}$$

where $J(0)$ is the value (height) of the performance surface when w is equal to $w(n)$ at iteration n.

To obtain a decaying equivalent expression, we introduce the relation

$$(1-2\mu)^{2n} = e^{-\frac{n}{\tau}} \tag{6.1.26}$$

where is the *time constant*. Under the condition $2\mu<<1$, we can use the approximation $\ln(1-2\mu) \cong -2\mu$. Therefore, the time constant has the value

$$\tau \cong \frac{1}{4\mu} \tag{6.1.27}$$

The above equation shows that Newton's algorithm has only *one mode of convergence* (one time constant).

6.2 Steepest-descent algorithm

To find the minimum value of the MSE surface, J_{min}, using the steepest descent algorithm, we proceed as follows: a) We start with the initial value $\mathbf{w}(0)$, usually using the null vector. b) At the MSE surface point that corresponds to $\mathbf{w}(0)$, we compute the gradient vector $\nabla J(\mathbf{w}(0))$. c) We compute the value $-\mu\nabla J(\mathbf{w}(0))$ and add it to $\mathbf{w}(0)$ to obtain $\mathbf{w}(1)$. d) We go back to step b) and continue the procedure until we find the optimum value of the vector coefficients.

If $\mathbf{w}(n)$ is the filter-coefficient vector at step n (time), then its updated value $\mathbf{w}(n+1)$ is given by (see also (6.1.3))

$$\mathbf{w}(n+1) = \mathbf{w}(n) - \mu\nabla J(\mathbf{w}(n)) \tag{6.2.1}$$

The gradient vector is equal to (see (6.1.17))

$$\nabla J(\mathbf{w}(n)) = -2\mathbf{p}_{dx} + 2\mathbf{R}_x\mathbf{w}(n) \tag{6.2.2}$$

and, hence, (6.2.1) becomes

$$\begin{aligned}\mathbf{w}(n+1) &= \mathbf{w}(n) + 2\mu[\mathbf{p}_{dx} - \mathbf{R}_x\mathbf{w}(n)] = \mathbf{w}(n) + \mu'[\mathbf{p}_{dx} - \mathbf{R}_x\mathbf{w}(n)] \\ &= [\mathbf{I} - \mu'\mathbf{R}_x]\mathbf{w}(n) + \mu'\mathbf{p}_{dx}\end{aligned} \tag{6.2.3}$$

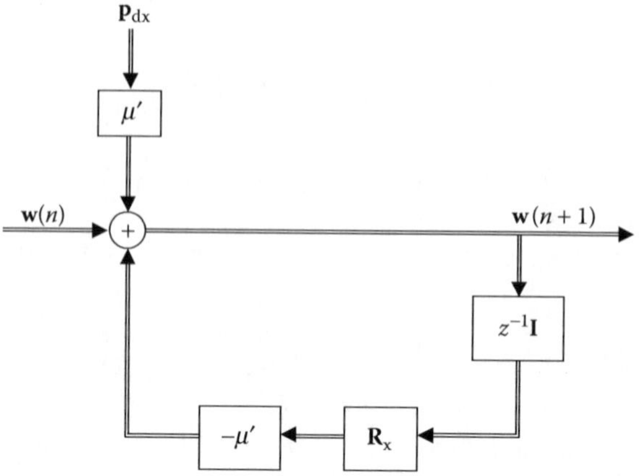

Figure 6.2.1 Signal-flow-graph representation of the steepest-descent algorithm.

where we set $\mu' = 2\mu$, and **I** is the identity matrix. The value of the primed step-size parameter must be much less than 1/2 for convergence. The signal-flow-graph of (6.2.3) is shown in Figure 6.2.1.

To apply the method of steepest descent, we must find first estimates of the autocorrelation matrix $\mathbf{R}_x$ and the cross-correlation vector $\mathbf{p}_{dx}$ from the data. This is necessary, since we do not have an ensemble of data to find $\mathbf{R}_x$ and $\mathbf{p}_{dx}$.

Stability (convergence) of the algorithm

Let

$$\xi(n) = \mathbf{w}(n) - \mathbf{w}^o \tag{6.2.4}$$

be the difference between the filter-coefficient vector and its optimum Wiener value $\mathbf{w}^o$. Next we write the first part of (6.2.3) in the form

$$\mathbf{w}(n+1) - \mathbf{w}^o = \mathbf{w}(n) - \mathbf{w}^o + \mu'[\mathbf{R}_x\mathbf{w}^o - \mathbf{R}_x\mathbf{w}(n)] \quad \text{or} \quad \xi(n+1) = [\mathbf{I} - \mu'\mathbf{R}_x]\xi(n) \tag{6.2.5}$$

But since $\mathbf{R}_x = \mathbf{Q}\Lambda\mathbf{Q}^T$ (see Table 5.1.1) and $\mathbf{I} = \mathbf{Q}\mathbf{Q}^T$, (6.2.5) becomes:

$$\begin{aligned}
&\xi(n+1) = [\mathbf{I} - \mu'\mathbf{Q}\Lambda\mathbf{Q}^T]\xi(n) \\
&or \\
&\mathbf{Q}^T\xi(\mathbf{n}+1) = [\mathbf{I} - \mu'\Lambda]\mathbf{Q}^T\xi(n) \qquad (6.2.6) \\
&or \\
&\xi'(n+1) = [\mathbf{I} - \mu'\Lambda]\xi'(n)
\end{aligned}$$

where the coordinate axis ξ' 's are orthogonal to ellipsoids (see Chapter 5), and is a diagonal matrix with its diagonal elements equal to the eigenvalues of $\mathbf{R}_x$. The kth row of (6.2.6), which represents the kth *natural mode* of the steepest descent algorithm, is (see Problem 6.2.1)

$$\xi'_k(n+1) = [1-\mu'\lambda_k]\xi'_k(n) \tag{6.2.7}$$

The above equation is a homogeneous difference equation, which has the following solution (see also Problem 6.1.1)

$$\xi'_k(n) = (1-\mu'\lambda_k)^n \xi'_k \qquad k = 0,1,\cdots,M-1, \quad n = 1,2,3,\cdots \tag{6.2.8}$$

For the above equation to converge (to be stable) as $n \to \infty$, we must set

$$-1 < 1-\mu'\lambda_k < 1 \qquad k = 0,1,2,\cdots,M-1 \tag{6.2.9}$$

From the right side of the inequality we obtain $-\mu'\lambda_k < 0$ *or* $\mu' > 0$, since λ_k are real and positive. From the left side of the inequality we obtain $\mu'\lambda_k < 2$ or $\mu' < 2/\lambda_k$, and hence

$$0 < \mu' < 2/\lambda_k \quad k = 0,1,\cdots,M-1 \quad or \quad 0 < \mu < 1/\lambda_k \quad k = 0,1,\cdots,M-1 \tag{6.2.10}$$

Under the above conditions and as $n \to \infty$, (6.2.8) becomes $\lim_{n\to\infty} \xi'_k(n) = 0$ or $\xi = 0$ since $\mathbf{Q}^T \neq 0$ and, thus, $\mathbf{w}(\infty) = \mathbf{w}^o$. Since (6.2.8) decays exponentially to zero, there exists a time constant that depends on the value of μ' and the eigenvalues of $\mathbf{R}_x$. Furthermore, (6.2.8) implies that immaterially of the initial value $\xi(0)$, $\mathbf{w}(n)$ always converges to $\mathbf{w}^o$ provided, of course, that (6.2.10) is satisfied. This is an important property of the steepest-descent algorithm. Since each row of (6.2.7) must decay as $n \to \infty$, it is necessary and sufficient that μ' obeys the following relationship

$$0 < \mu' < \frac{2}{\lambda_{max}} \tag{6.2.11}$$

Since $(1-\mu'\lambda_k)^n$ decays exponentially, there exists an exponential function with time constant τ_k such that $(e^{-1/\tau_k})^n = (1-\mu'\lambda_k)^n$ or $1-\mu'\lambda_k = e^{-1/\tau_k} = 1-\frac{1}{\tau_k}+\frac{1}{2!\tau_k^2}-\cdots$ Therefore, for small μ' and λ_k (large τ_k), we have

$$\tau_k \cong \frac{1}{\mu'\lambda_k} \qquad \mu'\lambda_k \ll 1 \tag{6.2.12}$$

In general, the kth time constant τ_k can be expressed in the form

$$\tau_k = -\frac{1}{\ln(1-\mu'\lambda_k)} \tag{6.2.13}$$

Transient behavior of MSE

Using (5.2.7) at time n we obtain the relation

$$J(\mathbf{w}(n)) = J_{\min} + \xi'^T \Lambda \xi' = J_{\min} + \sum_{k=0}^{M-1} \lambda_k \xi_k'^2(n) \tag{6.2.14}$$

Substituting the solution for $\xi_k'(n)$ from (6.2.8) in (6.2.14), we find the relation

$$J(\mathbf{w}(n)) = J_{\min} + \sum_{k=0}^{M-1} \lambda_k (1-\mu'\lambda_k)^{2n} \xi'^2(0) \tag{6.2.15}$$

It is obvious from the above equation (the factor in parenthesis is assumed to be less than one) that

$$\lim_{n\to\infty} J(\mathbf{w}(n)) = J_{\min} \tag{6.2.16}$$

From (6.2.14) we observe that the learning curve (plot of $J(\mathbf{w}(n))$ vs. n) consists of a sum of exponentials, each one corresponds to a natural mode of the algorithm.

Solution of the vector difference equation

If we set $n = 0$ in (6.2.3), we obtain

$$\mathbf{w}(1) = [\mathbf{I} - \mu'\mathbf{R_x}]\mathbf{w}(0) + \mu'\mathbf{p_{dx}} \tag{6.2.17}$$

Similarly, if we set $n = 1$, we find

$$\mathbf{w}(2) = [\mathbf{I} - \mu'\mathbf{R_x}]\mathbf{w}(1) + \mu'\mathbf{p_{dx}} \tag{6.2.18}$$

If we substitute (6.2.17) in (6.2.18) we obtain

$$\begin{aligned}\mathbf{w}(2) &= [\mathbf{I} - \mu'\mathbf{R}]^2\mathbf{w}(0) + \mu'\mathbf{p_{dx}} + [\mathbf{I} - \mu'\mathbf{R_x}]\mu'\mathbf{p_{dx}} \\ &= [\mathbf{I} - \mu'\mathbf{R_x}]^2\mathbf{w}(0) + \left(\sum_{j=0}^{1}[\mathbf{I} - \mu'\mathbf{R_x}]^j\right)\mu'\mathbf{p_{dx}}\end{aligned} \tag{6.2.19}$$

Therefore, at the nth step we obtain

$$\mathbf{w}(n) = [\mathbf{I} - \mu'\mathbf{R_x}]^n\mathbf{w}(0) + \left(\sum_{j=0}^{n-1}[\mathbf{I} - \mu'\mathbf{R_x}]^j\right)\mu'\mathbf{p_{dx}} \tag{6.2.20}$$

The above equation does not provide us with a way to study the convergence of $\mathbf{w}(n)$ to $\mathbf{w}^o$ as $n \to \infty$. We must decouple the equations and describe them in a different coordinate system. To accomplish this task we translate and then rotate the coordinate system.

After finding the eigenvalues and eigenvectors of $\mathbf{R}_x$, we create a diagonal matrix consisting of the eigenvalues of $\mathbf{R}_x$ and a matrix $\mathbf{Q}$ made up of the eigenvectors of $\mathbf{R}_x$ (see Chapter 5). Since $\mathbf{Q}^T\mathbf{Q} = \mathbf{Q}\mathbf{Q}^T = \mathbf{I}$, (6.2.3) takes the form

$$\mathbf{w}(n+1) = \mathbf{Q}[\mathbf{I} - \mu'\Lambda]\mathbf{Q}^T\mathbf{w}(n) + \mu'\mathbf{p}_{dx} \tag{6.2.21}$$

To uncouple the weights (coefficients), we multiply both sides of the above equation by $\mathbf{Q}^T$. Hence,

$$\mathbf{w}'(n+1) = [\mathbf{I} - \mu'\Lambda]\mathbf{w}'(n) + \mu'\mathbf{p}'_{dx} \tag{6.2.22}$$

where we define the following quantities

$$\mathbf{w}'(n+1) = \mathbf{Q}^T\mathbf{w}(n+1),\ \mathbf{w}'(n) = \mathbf{Q}^T\mathbf{w}(n),\ \mathbf{p}'_{dx} = \mathbf{Q}^T\mathbf{p}_{dx},\ \mathbf{w}^{o\prime} = \mathbf{Q}^T\mathbf{w}^o \tag{6.2.23}$$

Next, we obtain the relation

$$\begin{aligned}\mathbf{w}^{o\prime} &= \mathbf{Q}^T\mathbf{w}^o = \mathbf{Q}^T\mathbf{R}_x^{-1}\mathbf{p}_{dx} = \mathbf{Q}^T(\mathbf{Q}\Lambda\mathbf{Q}^T)^{-1}\mathbf{p}_{dx} \\ &= \mathbf{Q}^T\mathbf{Q}\Lambda^{-1}\mathbf{Q}^T\mathbf{p}_{dx} = \Lambda^{-1}\mathbf{p}'_{dx}\end{aligned} \tag{6.2.24}$$

since $\mathbf{Q}^{-1} = \mathbf{Q}^T$ and $(\mathbf{Q}\Lambda\mathbf{Q}^T)^{-1} = (\Lambda\mathbf{Q}^T)^{-1}\mathbf{Q}^{-1} = (\mathbf{Q}^T)^{-1}\Lambda^{-1}\mathbf{Q}^T = \mathbf{Q}\Lambda^{-1}\mathbf{Q}^T$.

The ith equation of the system given by (1.2.21) is

$$w'_i(n+1) = [1 - \mu'\lambda_i]w'_i(n) + \mu'p'_{idx} \qquad 0 \le i \le M-1 \tag{6.2.25}$$

By iteration, the above equation has the following solution (see Problem 6.2.2):

$$w'_i(n) = (1-\mu'\lambda_i)^n w'_i(0) + \mu'p'_{idx}\left(\sum_{j=0}^{n-1}(1-\mu'\lambda_i)^j\right) \tag{6.2.26}$$

If we set $\alpha_i = 1 - \mu'\lambda_i$, then (6.2.26) becomes

$$w'_i(n) = \alpha_i^n w'_i(0) + \mu'p'_{idx}\sum_{j=0}^{n-1}\alpha_i^j = \alpha_i^n w'_i(0) + \mu'p'_{idx}\frac{1-\alpha_i^n}{1-\alpha_i} \tag{6.2.27}$$

$$w'_i(n) = \mu'p'_{idx}\frac{1-\alpha_i^n}{1-\alpha_i} = \frac{1}{\lambda_i}[1-(1-\mu'\lambda_i)^n] \tag{6.2.28}$$

since the sum is a finite geometric series.

Example 6.2.1: For the development of this example we used the help of MATLAB. Hence, the data were found using x = randn(1,10), and the desired ones were found using d = conv([1 0.2],x) or d = filter([1 0.2],1,x). The correlation matrix was found using the function R = toeplitz(xc(1,10:11)), where xc = xcorr(x,'biased'). Similarly, we found the pdx cross-correlation. Hence,

$$\mathbf{R}=\begin{bmatrix}0.7346 & 0.0269\\ 0.0269 & 0.7346\end{bmatrix};\quad [\mathbf{Q},\mathbf{\Lambda}]=eig(R);$$

$$\mathbf{Q}=\begin{bmatrix}-0.7071 & 0.7071\\ 0.7071 & 0.7071\end{bmatrix};\quad \mathbf{\Lambda}=\begin{bmatrix}0.7077 & 0\\ 0 & 0.7071\end{bmatrix}$$

and $\mu' < 2/0.7071 = 2.8285$ for the solution to converge. We next choose $[w_0\ w_1]' = [0\ 0]'$ and hence $\mathbf{w}'(0)=\mathbf{Q}^{\mathrm{T}}\mathbf{w}(0)=\mathbf{0}$. Therefore, (6.2.27) becomes

$$w_i'(n)=\mu' p_{idx}'\frac{1-\alpha_i^n}{1-\alpha_i}=\frac{1}{\lambda_i}p_{idx}'[1-(1-\mu'\lambda_i)^n]$$

From (6.2.23) we obtain

$$\mathbf{p}_{\mathbf{dx}}'=\mathbf{Q}^{\mathrm{T}}\mathbf{p}_{\mathbf{dx}}=\begin{bmatrix}-0.7071 & 0.7071\\ 0.7071 & 0.7071\end{bmatrix}\begin{bmatrix}0.7399\\ -0.0003\end{bmatrix}=\begin{bmatrix}-0.5234\\ 0.5230\end{bmatrix}.$$

Therefore, the system is (we set $\mu' = 3$ for convergence)

$$w_0'(n)=\frac{1}{0.7077}(-0.5234)[1-(1-2\times 0.7071)^n]$$

$$w_1'(n)=\frac{1}{0.7614}0.5230[1-(1-2\times 0.7614)^n]$$

Since $\mathbf{w}'(n)=\mathbf{Q}^{\mathrm{T}}\mathbf{w}(n)$ and $\mathbf{Q}^{\mathrm{T}}=\mathbf{Q}^{-1}$ we find $\mathbf{w}(n)=\mathbf{Q}^{\mathrm{T}}\mathbf{w}'(n)$ and at the limit value when n approaches infinity, the filter coefficients take the values

$$\begin{bmatrix}w_0^o\\ w_1^o\end{bmatrix}=\mathbf{Q}^{\mathrm{T}}\mathbf{w}'=\begin{bmatrix}-0.7071 & 0.7071\\ 0.7071 & 0.7071\end{bmatrix}\begin{bmatrix}\dfrac{-0.5234}{0.7077}\\ \dfrac{0.5230}{0.7614}\end{bmatrix}=\begin{bmatrix}1.0087\\ -0.0373\end{bmatrix}$$

Problems

6.1.1 Verify (6.1.5).

6.1.2 Plot (6.1.5) for the following positive values of μ: (a) $0<\mu<\frac{1}{2r_x(0)}$ (b) $\mu\cong\frac{1}{2r_x(0)}$ (c) $\frac{1}{2r_x(0)}<\mu<\frac{1}{r_x(0)}$ with $r_x(0) = 1$. Identify the type of your plots.

6.1.3 Verify (6.1.25).

6.1.4 Using Newton's algorithm, find the third root of 8. Start with $\mathbf{w}(0) = 1.2$.

6.1.5 Let the MSE is given by the non-quadratic equation $J = 2 - \frac{1}{35}[(1 - w^2)(4.5 + 3.5w)]$. Find and plot the recursive Newton's algorithm for the coefficient $w(n)$, $n = 0, 1, 2, \ldots$ Start with $w(0) = 2$.

6.2.1 Verify (6.2.7).

6.2.2 Verify (6.2.26).

6.2.3 Let $\hat{\mathbf{R}}_x = \begin{bmatrix} 1 & 0.7 \\ 0.7 & 1 \end{bmatrix}$ and $\hat{\mathbf{p}}_{dx} = [0.7 \quad 0.5]^T$ are the estimates derived from the data. Find the vector $\mathbf{w}(n)$.

6.2.4 Find an equivalent expression of $J(\mathbf{w}(n)) = J_{\min} + \xi'(n)\Lambda\xi'(n)$ as a function of $\mathbf{w}(n)$, $\mathbf{p}_{dx}(n)$, and $\mathbf{R}_x$.

6.2.5 The correlation matrix of the filter input is $\mathbf{R} = \begin{bmatrix} 1 & 0.85 \\ 0.85 & 1 \end{bmatrix}$ with eigenvalues $\lambda_0 = 1.85$ and $\lambda_1 = 0.15$. Plot the learning curve in a semi-log format and find the two time constants. Compare the results obtained from the graph and those given analytically (see (1.2.12)).

6.2.6 Plot $|1 - \mu'\lambda_{\min}|$ and $|1 - \mu'\lambda_{\max}|$ vs. μ' for two hypothetical values of $\lambda_{\min} = 0.4$ and $\lambda_{\max} = 0.8$. Find the optimum value μ'_{opt}, which is given by the intersection of the two curves. All the other values of the eigenvalues create lines between the above two. When $\mu' = \mu'_{opt}$, the speed of convergence of the steepest-descent algorithm is determined by the factor $\alpha = 1 - 2\mu'_{opt}\lambda_{\min}$.

Hints-solutions-suggestions

6.1.1:

$$w(1) = aw(0) + bw^o \,(a = 1 - 2\mu r(0), b = 2\mu r(0)), w(2) = aw(1) + bw^o$$

$$= a(aw(0) + bw^o) + bw^o$$

$$= a^2 w(0) + abw^o + bw^o \Rightarrow w(n) = a^n w(0) + (a^{n-1} + a^{n-2} + \cdots + 1)bw^o$$

$$= a^n w(0) + \frac{1 - a^n}{1 - a} bw^o,$$

but $1 - a = b \Rightarrow w(n) = a^n w(0) + w^o - a^n w^o = w^o + (w(0) - w^o)a^n$.

6.1.2:
Answer: (a) over-damped case. (b) critically damped. (c) under-damped case.

6.1.3:

$$\begin{aligned} J(n) &= J_{\min} + \xi^T \mathbf{R}_{\mathbf{x}} \xi = J_{\min} + (1-2\mu)^n \xi^T(0) \mathbf{R}_{\mathbf{x}} (1-2\mu)^n \xi(0) \\ &= J_{\min} + (1-2\mu)^{2n} \xi^T(0) \mathbf{R}_{\mathbf{x}} \xi(0) \end{aligned}$$

where (5.2.7) and (6.1.24) were used. But $\xi^T(0)\mathbf{R}_{\mathbf{x}}\xi(0) = J(0) - J_{\min}$ and Problem 6.1.3 is proved.

6.1.4:
Hint: $f(x) = x^3 - 8$, hence $x(n+1) = x(n) - [x^3(n) - 8]/[3x^2(n)]$

6.1.5:

$$w(n+1) = w(n) - [(\partial J(w)/\partial w)/(\partial^2 J(w)/\partial w^2)] = w(n) - \frac{10.5w^2(n) + 9w(n) - 3.5}{21w(n) + 9}$$

6.2.1:

$$\begin{bmatrix} \xi'_0(n+1) \\ \xi'_1(n+1) \\ \vdots \\ \xi'_{M-1}(n+1) \end{bmatrix} = \left(\begin{bmatrix} 1 & 0 & \cdots & 0 \\ 0 & 1 & \cdots & 0 \\ & \vdots & & \\ 0 & 0 & \cdots & 1 \end{bmatrix} - \mu \begin{bmatrix} \lambda_0 & 0 & \cdots & 0 \\ 0 & \lambda_1 & \cdots & 0 \\ & \vdots & & \\ 0 & 0 & \cdots & \lambda_{M-1} \end{bmatrix} \right) \begin{bmatrix} \xi'_0(n) \\ \xi'_1(n) \\ \vdots \\ \xi'_{M-1} \end{bmatrix} = \begin{bmatrix} (1-\mu\lambda_0)\xi'_0(n) \\ (1-\mu\lambda_1)\xi'_1(n) \\ \vdots \\ (1-\mu\lambda_{M-1})\xi'_{M-1}(n) \end{bmatrix}$$

and the results are found since the equality of two vectors means equality of each corresponding element.

6.2.2:

$$\begin{aligned} &\text{Start with } n = 0 \Rightarrow w'_i(1) = (1-\mu'\lambda_i)^2 w'_i(0) + \mu' p'_{idx}, \text{ next set } n = 1 \Rightarrow w'_i(2) \\ &= (1-\mu'\lambda_i)[(1-\mu'\lambda_i)w'_i(0) + \mu'\mathrm{p}'_{idx}] + \mu'\mathrm{p}'_{idx} \\ &= (1-\mu'\lambda_i)^2 w'_i(0) + \mu'\mathrm{p}'_{idx}[1 + (1-\mu'\lambda_i)] \\ &= (1-\mu'\lambda_i)^2 w'_i(0) + \mu' p'_{idx} \left(\sum_{j=0}^{2-1} (1-\mu'\lambda_i)^j \right) \end{aligned}$$

and, therefore, at the nth iteration we obtain (1.2.26).

6.2.3:
Using the MATLAB function [v,d] = eig(R), we obtain the following eigenvalues and eigenvectors: $\lambda_1 = 0.3000$, $\lambda_2 = 1.7000$, $\mathbf{q}_1 = [-0.7071\ \ 0.7071]^T$, $\mathbf{q}_2 = [0.7071\ \ 0.7071]^T$. The step-size factor must be set equal to: $\mu' < 2/1.7 = 1.1765$,

so that the solution converges. Choosing $\mathbf{w}(0) = \mathbf{0}$, implies that $\mathbf{w}'(0) = \mathbf{Q}^T\mathbf{w}(0) = \mathbf{0} \Rightarrow$

$$\mathbf{p}'_{dx} = \mathbf{Q}^T\mathbf{p}_{dx} = \begin{bmatrix} -0.7071 & 0.7071 \\ 0.7071 & 0.7071 \end{bmatrix}\begin{bmatrix} 0.7 \\ 0.5 \end{bmatrix} = \begin{bmatrix} -0.1414 \\ 0.8485 \end{bmatrix}$$

$$w'_0(n) = \frac{1}{0.3}(-0.1414)[1-(1-\mu'0.3)^n];\, w'_1(0) = \frac{1}{1.7}0.8485[1-(1-\mu'1.7)^n].$$

Since

$$\mathbf{w}'(n) = \mathbf{Q}^T\mathbf{w}(n),\ \mathbf{Q}^T = \mathbf{Q}^{-1},\ \mathbf{w}(n) = \mathbf{Q}^T\mathbf{w}'(n), \text{ we find}$$

$$\begin{bmatrix} w_0(n) \\ w_1(n) \end{bmatrix} = \begin{bmatrix} -0.7071 & 0.7071 \\ 0.7071 & 0.7071 \end{bmatrix}\begin{bmatrix} \frac{1}{0.3}(-0.1414)[1-(1-\mu'0.3)^n] \\ \frac{1}{1.7}0.8485[1-(1-\mu'1.7)^n] \end{bmatrix}$$

$$= \begin{bmatrix} \frac{-0.7071}{0.3}(-0.1414)[1-(1-\mu'0.3)^n] + \frac{0.7071}{1.7}0.8485[1-(1-\mu'1.7)^n] \\ \frac{0.7071}{0.3}(-0.1414)[1-(1-\mu'0.3)^n] + \frac{0.7071}{1.7}0.8485[1-(1-\mu'1.7)^n] \end{bmatrix}$$

6.2.4:

$$j(\mathbf{w}(n)) = \sigma_d^2 - \mathbf{p}_{dx}^T\mathbf{w}^o + \xi(n)\mathbf{Q}\mathbf{Q}^T\mathbf{R}_X\mathbf{Q}\mathbf{Q}^T\xi(n)$$
$$= \sigma_d^2 - \mathbf{p}_{dx}^T\mathbf{w}^o + (\mathbf{w}(n) - \mathbf{w}^o)\mathbf{R}_X(\mathbf{w}(n) - \mathbf{w}^o)$$
$$= \sigma_d^2 - \mathbf{p}_{dx}^T\mathbf{w}^o + [\mathbf{w}^T(n) - \mathbf{w}^o]\mathbf{R}_X[\mathbf{w}(n) - \mathbf{w}^o]$$
$$= \sigma_d^2 - \mathbf{p}_{dx}^T\mathbf{w}^o + \mathbf{w}^{oT}\mathbf{R}_X\mathbf{w}(n) - \mathbf{w}^{oT}\mathbf{R}_X\mathbf{w}(n) - \mathbf{w}^T(n)\mathbf{R}_X\mathbf{w}^o$$
$$+ \mathbf{w}^{oT}\mathbf{R}_X\mathbf{w}^o.$$

But $(\mathbf{R}_X\mathbf{w}^o)^T = \mathbf{w}^{oT}\mathbf{R}_x^T = \mathbf{w}^{oT}\mathbf{R}_X = \mathbf{p}_{dx}^T, \mathbf{R}_X\mathbf{w}^o = \mathbf{p}_{dx}, \mathbf{p}_{dx}^T\mathbf{w}^o$

$= \mathbf{w}^{oT}\mathbf{p}_{dx}$ and, hence,

$$J(\mathbf{w}(n)) = \sigma_d^2 - 2\mathbf{w}^T(n)\mathbf{p}_{dx} + \mathbf{w}^T(n)\mathbf{R}_X\mathbf{w}(n).$$

6.2.5:
Hint: The time constants are found from the two slopes of the graph.

6.2.6:
Hint: $\mu'_{opt} = 2/[\lambda_{min} + \lambda_{max}],\ \alpha = (\lambda_{max}/\lambda_{min} - 1)/(\lambda_{max}/\lambda_{min} + 1)$

chapter 7

The least mean-square (LMS) algorithm

7.1 Introduction

In this chapter, we present the celebrated *least mean-square* (LMS) *algorithm*, developed by Widrow and Hoff in 1960. This algorithm is a member of stochastic gradient algorithms, and because of its robustness and low computational complexity, it has been used in a wide spectrum of applications.

The LMS algorithm has the following most important properties:

1. It can be used to solve the Wiener–Hopf equation without finding matrix inversion. Furthermore, it does not require the availability of the autocorrelation matrix of the filter input and the cross correlation between the filter input and its desired signal.
2. Its form is simple as well as its implementation, yet it is capable of delivering high performance during the adaptation process.
3. Its iterative procedure involves: a) computing the output of an FIR filter produced by a set of tap inputs (filter coefficients), b) generation of an estimated error by comparing the output of the filter to a desired response, and c) adjusting the tap weights (filter coefficients) based on the estimation error.
4. The correlation term needed to find the values of the coefficients at the $n + 1$ iteration contains the stochastic product $\mathbf{x}(n)e(n)$ without the expectation operation that is present in the steepest-descent method.
5. Since the expectation operation is not present, each coefficient goes through sharp variations (noise) during the iteration process. Therefore, instead of terminating at the Wiener solution, the LMS algorithm suffers random variation around the minimum point (optimum value) of the error-performance surface.
6. It includes a step-size parameter, μ, that must be selected properly to control stability and convergence speed of the algorithm.
7. It is stable and robust for a variety of signal conditions.

7.2 Derivation of the LMS algorithm

In Chapter 6 we developed the following relations using the steepest-descent method

$$\mathbf{w}(n+1)=\mathbf{w}(n)-\mu\nabla J(\mathbf{w}(n)) \tag{7.2.1}$$

$$\nabla J(\mathbf{w}(n))=-2\mathbf{p}_{\mathbf{dx}}+2\mathbf{R}_{\mathbf{x}}\mathbf{w}(n) \tag{7.2.2}$$

The simplest choices of estimators for $\mathbf{R}_{\mathbf{x}}$ and $\mathbf{p}_{\mathbf{dx}}$ are the *instantaneous estimates* defined by:

$$\mathbf{R}_{\mathbf{x}}\cong\mathbf{x}(n)\mathbf{x}^{\mathrm{T}}(n),\quad \mathbf{p}_{\mathbf{dx}}\cong d(n)\mathbf{x}(n) \tag{7.2.3}$$

Substituting the above values in (7.2.2) and then combining (7.2.1) and (7.2.2), we obtain

$$\begin{aligned}\mathbf{w}(n+1)&=\mathbf{w}(n)+2\mu\mathbf{x}(n)[d(n)-\mathbf{x}^{\mathrm{T}}(n)\mathbf{w}(n)]\\&=\mathbf{w}(n)+2\mu\mathbf{x}(n)[d(n)-\mathbf{w}^{\mathrm{T}}(n)\mathbf{x}(n)]\\&=\mathbf{w}(n)+2\mu e(n)\mathbf{x}(n)\end{aligned} \tag{7.2.4}$$

where

$$y(n)=\mathbf{w}^{\mathrm{T}}(n)\mathbf{x}(n)\quad \text{filter output} \tag{7.2.5}$$

$$e(n)=d(n)-y(n)\quad \text{error} \tag{7.2.6}$$

$$\mathbf{w}(n)=[w_0(n)\;\; w_1(n)\cdots w_{M-1}(n)]^{\mathrm{T}}\quad \text{filter taps at time } n \tag{7.2.7}$$

$$\mathbf{x}(n)=[x(n)\;\; x(n-1)\;\; x(n-2)\cdots x(n-M+1)]^{\mathrm{T}}\quad \text{input data} \tag{7.2.8}$$

The algorithm defined by (7.2.4), (7.2.5), and (7.2.6) constitute the adaptive LMS algorithm. The algorithm at each iteration requires that $\mathbf{x}$(n), $d(n)$, and $\mathbf{w}(n)$ are known. The LMS algorithm is a *stochastic gradient algorithm* if the input signal is a stochastic process. This results in varying the pointing direction of the coefficient vector during the iteration. An FIR adaptive filter realization is shown in Figure 7.2.1. Figure 7.2.2 shows the block-diagram representation of the LMS filter. Table 7.2.1 presents the LMS algorithm.

Book LMS MATLAB function

```
function[w,y,e,J]=aalms(x,dn,mu,M)
%function[w,y,e,J]=aalms(x,dn,mu,M);
```

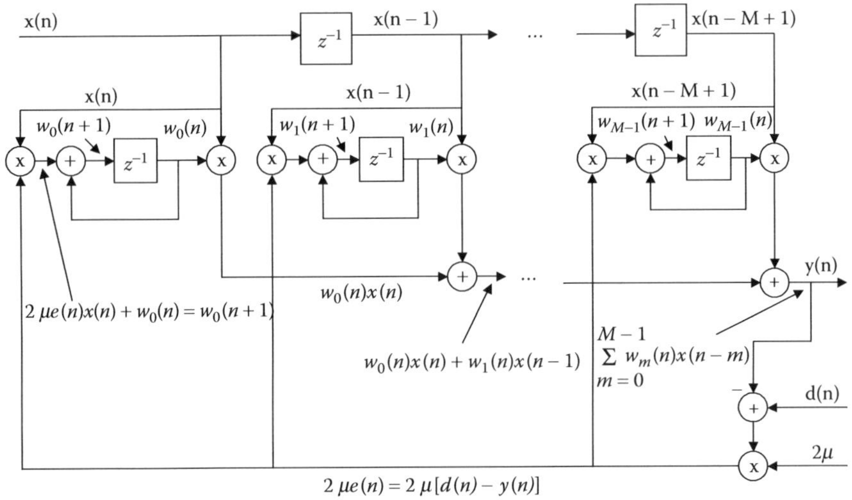

Figure 7.2.1 FIR LMS filter realization.

```
%all quantities are real-valued;
%x=input data to the filter; dn=desired signal;
%M=order of the filter;
%mu=step-size factor; x and dn must be of the same
%length;
N=length(x);
y=zeros(1,N); %initialized output of the filter;
w=zeros(1,M); %initialized filter coefficient vector;
for n=M:N
        x1=x(n:-1:n-M+1); %for each n the vector x1 is
```

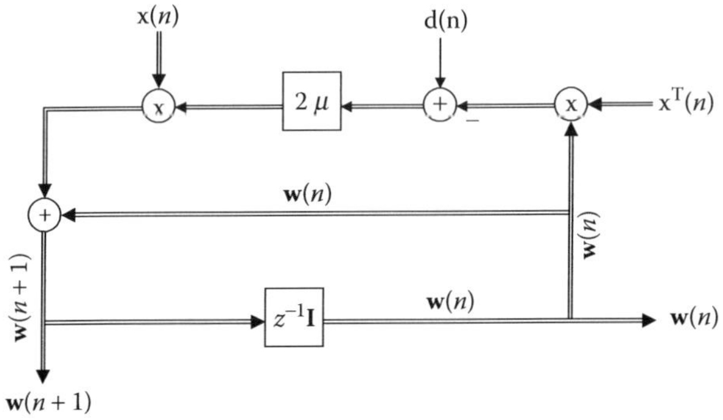

Figure 7.2.2 Block diagram representation of the LMS algorithm.

```
% of length M with elements from x in reverse order;
        y(n)=w*x1';
        e(n)=dn(n)-y(n);
        w=w+2*mu*e(n)*x1;
end;
J=e.^2;%J is the learning curve of the adaptation;
```

The book MATLAB function aalms1.m provides the changing values of the filter coefficients as a function of the iteration number n. The learning curve is easily found using the MATLAB command: **J = e.^2;**

Book LMS MATLAB function to provide changing values

```
function[w,y,e,J,w1]=aalms1(x,dn,mu,M)
%function[w,y,e,J,w1]=aalms1(x,dn,mu,M);
%this function provides also the changes of two filter
%coefficients versus iterations;
%all quantities are real-valued;
%x=input data to the filter; dn=desired signal;
%M=order of the filter;
%mu=step-size; x and dn must be of the same length;
%each column of the matrix w1 contains the history
%of each filter coefficient;
N=length(x);
y=zeros(1,N);
w=zeros(1,M); %initialized filter coefficient vector;
for n=M:N
x1=x(n:-1:n-M+1); %for each n the vector x1 of
%length M is produced with elements
%from x in reverse order;
        y(n)=w*x1';
        e(n)=dn(n)-y(n);
        w=w+2*mu*e(n)*x1;
        w1(n-M+1,:)=w(1,:);

end;

J=e.^2;%J is the learning curve of the adaptive
%process; each column of the matrix w1
%depicts the history of each filter coefficient;
```

7.3 Examples using the LMS algorithm

The following examples will elucidate the use of the LMS algorithm to different areas of engineering and will create an appreciation for the versatility of this important algorithm.

Table 7.2.1 The LMS Algorithm for an Mth-Order FIR Adaptive Filter

Inputs:	M = filter length
	μ = step-size factor
	$\mathbf{x}(n)$ = input data to the adaptive filter
	$\mathbf{w}(0)$ = initialization filter vector = $\mathbf{0}$
Outputs:	$y(n)$ = adaptive filter output = $\mathbf{w}^T(n)\mathbf{x}(n) \equiv \hat{d}(n)$
	$e(n) = d(n) - y(n)$ = error
	$\mathbf{w}(n+1) = \mathbf{w}(n) + 2\mu e(n)\mathbf{x}(n)$

Example 7.3.1 (Linear Prediction): We can use an adaptive LMS filter as a predictor as shown in Figure 7.3.1. The data $x(n)$ were created by passing a zero-mean white noise $v(n)$ through an autoregressive (AR) process described by the difference equation: $x(n) = 0.6010x(n-1) - 0.7225x(n-2) + v(n)$. The LMS filter is used to predict the values of the AR filter parameters 0.6010 and –0.7225. A two-coefficient LMS filter predicts $x(n)$ by

$$\hat{x}(n) = \sum_{i=0}^{1} w_i(n)x(n-1-i) \equiv y(n) \tag{7.3.1}$$

Figure 7.3.2 shows the trajectories of w_0 and w_1 vs. the number of iterations for two different values of step-size ($\mu = 0.02$ and $\mu = 0.005$). The

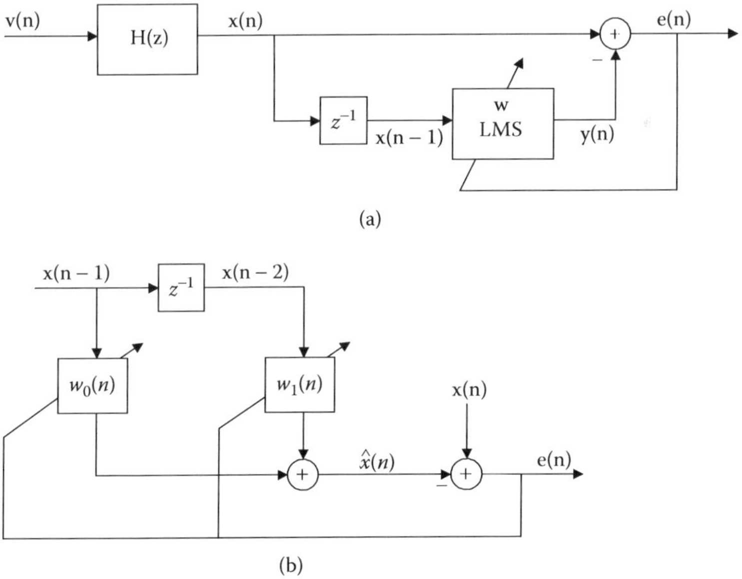

Figure 7.3.1 (a) Linear predictor LMS filter. (b) Two-coefficient adaptive filter with its adaptive weight-control mechanism.

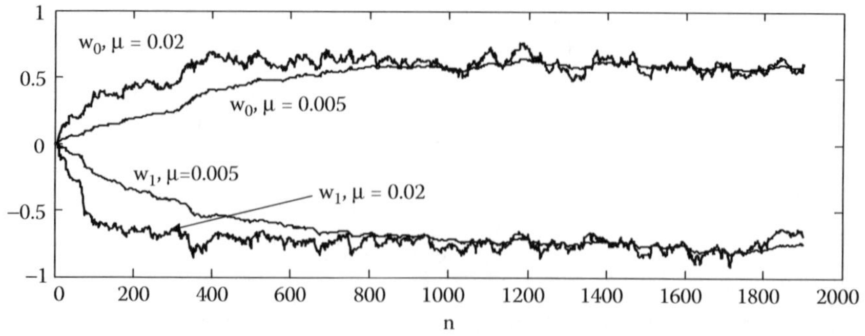

Figure 7.3.2 Convergence of two-element LMS adaptive filter used as linear predictor.

adaptive filter is a two-coefficient filter. The noise is white and Gaussian distributed. The figure shows fluctuations in the values of coefficients as they converge to a neighborhood of their optimum value, 0.6010 and 0.7225, respectively. As the step-size μ becomes smaller, the fluctuations are not as large, but the convergence speed to the optimal values is slower.

Book one-step LMS predictor MATLAB function

```
function[w,y,e,J,w1]=aalmsonesteppredictor(x,mu,M)
%function[w,J,w1]=aalmsonesteppredictor(x,mu,M);
%x=data=signal plus noise;mu=step size factor;M=number
%of filter
%coefficients;w1 is a matrix and each column is the
%history of each
%filter coefficient versus time n;
N=length(x);
y=zeros(1,N);
w=zeros(1,M);
for n=M:N-1
        x1=x(n:-1:n-M+1);
        y(n)=w*x1';
        e(n)=x(n+1)-y(n);
        w=w+2*mu*e(n)*x1;
        w1(n-M+1,:)=w(1,:);
end;
J=e.^2;
%J is the learning curve of the adaptive process;
```

Example 7.3.2 (Modeling): Adaptive filtering can also be used to find the coefficients of an unknown filter as shown in Figure 7.3.3. The data $x(n)$ were created similar to those in Example 7.3.1. The desired signal is given by $d(n) = x(n) - 2x(n-1) + 4x(n-2)$. If the output $y(n)$ is approximately

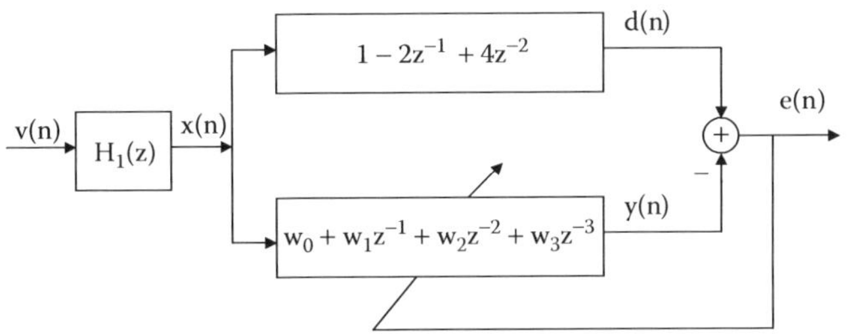

Figure 7.3.3 Identification of an unknown system.

equal to $d(n)$, it implies that the coefficients of the LMS filter are approximately equal to those of the unknown system. Figure 7.3.4 shows the ability of the LMS filter to identify the unknown system. After 500 iterations, the system is practically identified. For this example we used $\mu = 0.01$. It is observed that the fourth coefficient is zero, as it should be, since the system to be identified has only three coefficients and the rest are zero.

Example 7.3.3 (Noise Cancellation): A noise cancellation scheme is shown in Figure 7.3.5. In this example, we introduce the following values: $H_1(z) = 1$ (or $h(n) = \delta(n)$), $v_1(n)$ = white noise = $v(n)$, L = 1, $s(n) = \sin(0.2\pi n)$. Therefore, the input signal to the filter is $x(n) = s(n-1) + v(n-1)$ and the desired signal is $d(n) = s(n) + v(n)$.The Book LMS MATLAB program named aalms was used. Figure 7.3.6 shows the signal, the signal plus noise, and the outputs of the filter for two different sets of coefficients: $M = 4$ and $M = 12$.

Example 7.3.4 (Power Spectrum Approximation): If a stochastic process is the output of an autoregressive (AR) system when its input is a white noise with variance σ_v^2, e.g.,

$$x(n) = \sum_{k=1}^{N} a_k x(n-k) + v(n) \tag{7.3.2}$$

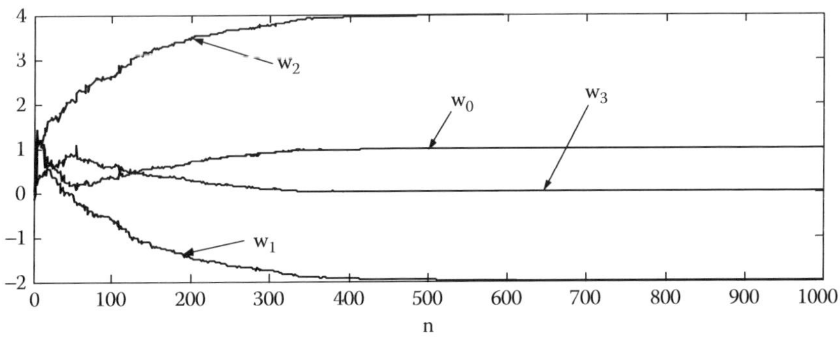

Figure 7.3.4 LMS adaptive filter coefficients for system modeling.

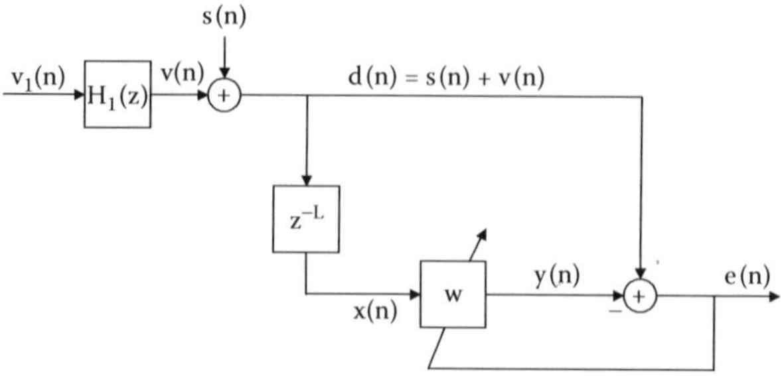

Figure 7.3.5 Adaptive LMS noise cancellation scheme.

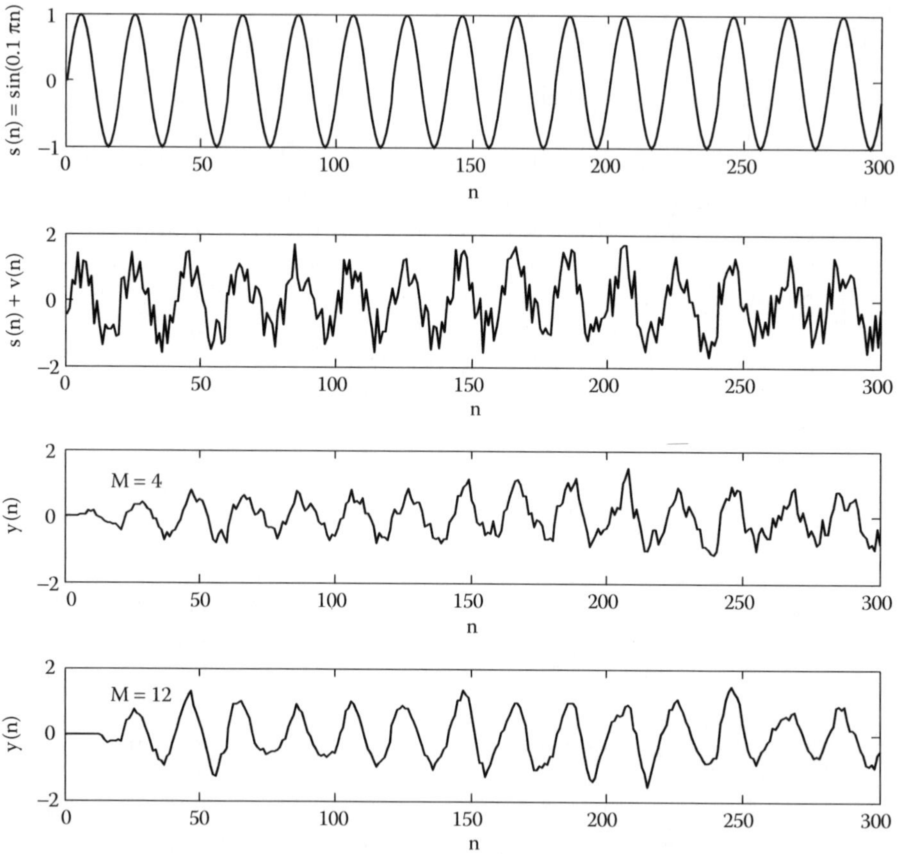

Figure 7.3.6 Noise cancellation of adaptive LMS filter.

the power spectrum corresponding to the stochastic process is given by (see Kay, 1993)

$$S_x(e^{j\omega}) = \frac{\sigma_v^2}{|A(e^{j\omega})|^2} \tag{7.3.3}$$

where

$$A(e^{j\omega}) = 1 - \sum_{k=1}^{N} a_k e^{-j\omega k} \tag{7.3.4}$$

Equation (7.3.2) is also written in the form

$$v(n) = x(n) - \sum_{k=1}^{N} a_k x(n-k) \tag{7.3.5}$$

where $v(n)$ is the non-predictable portion of the signal or the *innovations* of the AR process. Because $v(n)$ is the non-predictable portion of $x(n)$, it suggests to use an adaptive linear predictor for spectrum estimation. If the stochastic process is the result of an AR process, the LMS filter coefficients will be much closer to those of the AR system, and the two spectra will also be very close to each other.

The steps that closely approximate the coefficients of (7.3.2), using an LMS adaptive filter, are:

1. Use the adaptive LMS filter in the predictive mode (see Figure 7.3.7)
2. Average the K most recent values of $\hat{w}$
3. Compute the power spectrum

$$\hat{S}_x(e^{j\omega}) = 1\Big/\left|1 - \sum_{k=1}^{M} \hat{w}_k e^{-j\omega k}\right|^2 = 1\Big/\left|1 - \sum_{k=1}^{M} \hat{w}_k z^{-k}\right|^2_{z=e^{j\omega}} \tag{7.3.6}$$

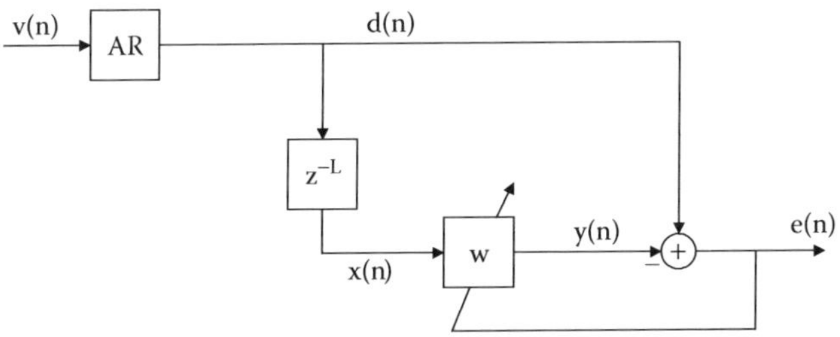

Figure 7.3.7 LMS adaptive filter for power spectrum estimation.

Let an exact stochastic process be created by an AR system having poles at $(z_1, z_2) = 0.95e^{\pm j\pi/4}$. To find the difference equation, which characterized the AR system, apply the definition of the system in the z-domain. Hence, we write

$$H(z) = \frac{output}{input} = \frac{X(z)}{\sigma_v^2 V(z)}$$

$$= \frac{1}{\left(1 - 0.95e^{j\frac{\pi}{4}}z^{-1}\right)\left(1 - 0.95e^{-j\frac{\pi}{4}}z^{-1}\right)} = \frac{1}{1 - 1.3435z^{-1} + 0.9025z^{-2}}$$

The above equation can be written as:

$$X(z) - 1.3435z^{-1}X(z) + 0.9025z^{-2}X(z) = \sigma_v^2 V(z)$$

Taking the inverse z-transform of both sides of the above equation, we obtain the difference equation describing the AR system given by

$$x(n) = 1.3435x(n-1) - 0.9025x(n-2) + \sigma_v^2 v(n) \tag{7.3.7}$$

The power spectrum is given by (see (7.3.3))

$$S_x(e^{j\omega}) = \frac{\sigma_v^2}{\left|1 - 1.3435e^{-j\omega} + 0.9025e^{-j2\omega}\right|^2} = \frac{\sigma_v^2}{\left|1 - 1.3435z^{-1} + 0.9025z^{-2}\right|^2_{z=e^{j\omega}}} \tag{7.3.8}$$

Figure 7.3.8 shows the true spectrum and the approximate one. We assumed that the desired signal was produced by the AR filter given by (7.3.7). The approximate spectrum was found using the following constants: $\mu = 0.02$, $M = 3$, $N = 1000$, avn = 3, $x(n) = \text{dn}(n-1)$, and $\sigma_v^2 = 0.3385$. The function aapowerspctraav1 will average the output **w** over a number of times as desired by the reader. If we had guessed $M = 2$ and avn = 5, the two curves will be approximately equal and the filter coefficients are also approximately equal: 1.3452, –0.8551.

Book MATLAB function to obtain power spectra

```
function[wa1,v]=aapowerspectraav1(a1,a2,a3,mu,M,N,avn,vr)
%aapowerspectraav1(a1,a2,a3,mu,M,N,avn,vr);
wa=zeros(1,M);
dn=zeros(1,N);x=zeros(1,N);
for k=1:avn
        for n=4:N
            v(n)=vr*(rand-.5);
```

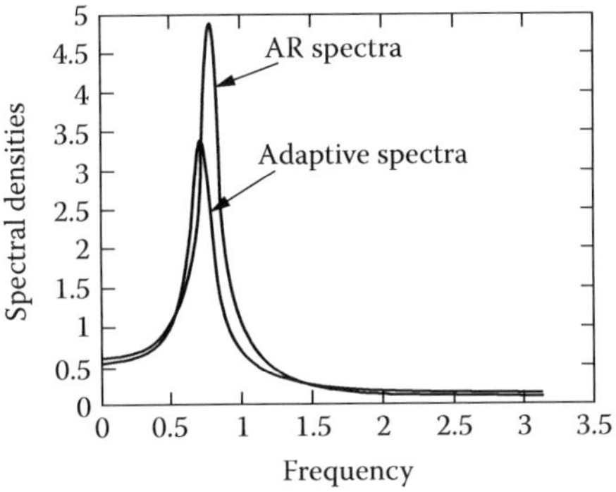

Figure 7.3.8 Power spectrum approximation using the LMS filter.

```
        dn(n)=-a1*dn(n-1)-a2*dn(n-2)-a3*dn(n-3)+v(n);
          x(n)=dn(n-1);
        end;
        [w]=aalms(x,dn,mu,M);
        wa=wa+w;
end;
wa1=wa/avn;
%this function gives averaged w's to be used for
%finding
%the approximate spectrum of the output of an AR filter
%up to the third order;M=number of LMS coefficients;
%N=length of desired signal and input signal to LMS
%filter;
%avn=number of times w's are averaged;the function is
%easily
%modified for AR filter with more coefficients;vr=con-
%trols the
%variance of the white noise, multiplies the quantity
%(rand-.5);
%the function up to the first end produces the output
%from
%the AR filter with coefficients a1,a2 and a3;
```

Plotting the spectra:

```
sx=freqz(var(v),[1 -[wa1]],512);
n=0:pi/512:pi-(pi/512);
plot(n,abs(sx)); xlabel('Radians');ylabel('Power...
spectrum');
%this function is useful up to three coefficients
%AR model;
%var controls the variance of the input noise;
```

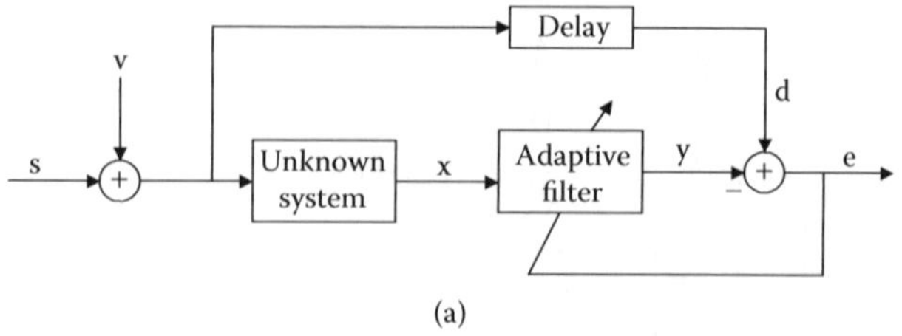

Figure 7.3.9(a) Creating an inverse system using an LMS filter.

Example 7.3.5 (Inverse System Identification): To find the inverse of an unknown filter, we place the adaptive filter in series with the unknown system as shown in Figure 7.3.9a. The delay is needed so that the system is causal. Figure 7.3.9b shows a typical learning curve. In this example we used four coefficients FIR filter, and the input to the unknown system was a sine function with a white Gaussian noise.

7.4 Performance analysis of the LMS algorithm

By subtracting the Wiener filter $\mathbf{w}^o$ (see (4.3.5)) from both sides of (7.2.4), we obtain the following equation

$$\mathbf{w}(n+1)-\mathbf{w}^o = \mathbf{w}(n)-\mathbf{w}^o + 2\mu e(n)\mathbf{x}(n) \tag{7.4.1}$$

The vectors $\xi(n+1)=\mathbf{w}(n+1)-\mathbf{w}^o$ and $\xi(n)=\mathbf{w}(n)-\mathbf{w}^o$ are known as the *weight-error* vectors, which are described on a coordinate system shifted by

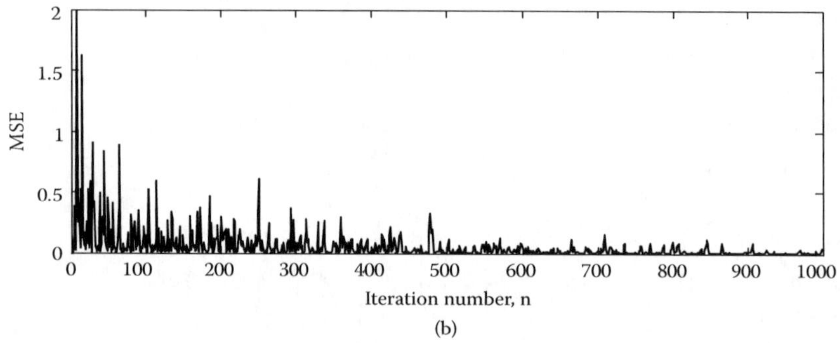

Figure 7.3.9(b) The learning curve of the inverse system problem.

$\mathbf{w}^o$ on the $\mathbf{w}$ plane. Therefore, (7.4.1) becomes

$$\xi(n+1) = \xi(n) + 2\mu e(n)\mathbf{x}(n) \tag{7.4.2}$$

$$\begin{aligned} e(n) &= d(n) - y(n) = d(n) - \mathbf{w}^{\mathrm{T}}(n)\mathbf{x}(n) = d(n) - \mathbf{x}^{\mathrm{T}}(n)\mathbf{w}(n) \\ &= d(n) - \mathbf{x}^{\mathrm{T}}(n)\mathbf{w}(n) - \mathbf{x}^{\mathrm{T}}(n)\mathbf{w}^o + \mathbf{x}^{\mathrm{T}}(n)\mathbf{w}^o \\ &= d(n) - \mathbf{x}^{\mathrm{T}}(n)\mathbf{w}^o - \mathbf{x}^{\mathrm{T}}(n)[\mathbf{w}(n) - \mathbf{w}^o] \\ &= e^o(n) - \mathbf{x}^{\mathrm{T}}(n)\xi(n) = e^o(n) - \xi^T(n)\mathbf{x}(n) \end{aligned} \tag{7.4.3}$$

where

$$e^o(n) = d(n) - \mathbf{x}^{\mathrm{T}}(n)\mathbf{w}^o \tag{7.4.4}$$

is the error when the filter is optimum. Substituting (7.4.3) in (7.4.2) and rearranging, we obtain

$$\begin{aligned} \xi(n+1) &= \xi(n) + 2\mu[e^o(n) - \mathbf{x}^{\mathrm{T}}(n)\xi(n)]\mathbf{x}(n) \\ &= \xi(n) + 2\mu\mathbf{x}(n)[e^o(n) - \mathbf{x}^{\mathrm{T}}(n)\xi(n)] \\ &= [\mathbf{I} - 2\mu\mathbf{x}(n)\mathbf{x}^{\mathrm{T}}(n)]\xi(n) + 2\mu e^o(n)\mathbf{x}(n) \end{aligned} \tag{7.4.5}$$

where $\mathbf{I}$ is the identity matrix and $e^o(n) - \mathbf{x}^{\mathrm{T}}(n)\xi(n)$ is a scalar. Next, we take the expectation of both sides of (7.4.5)

$$\begin{aligned} E\{\xi(n+1)\} &= E\{[\mathbf{I} - 2\mu\mathbf{x}(n)\mathbf{x}^{\mathrm{T}}(n)]\xi(n)\} + 2\mu E\{e^o(n)\mathbf{x}(n)\} \\ &= E\{[\mathbf{I} - 2\mu\mathbf{x}(n)\mathbf{x}^{\mathrm{T}}(n)]\xi(n)\} \end{aligned} \tag{7.4.6}$$

Since $e^o(n)$ is orthogonal to all data (see Section 4.3), the last expression is identically zero. The expression

$$E\{\mathbf{x}(n)\mathbf{x}^{\mathrm{T}}(n)\xi(n)\} = E\{\mathbf{x}(n)\mathbf{x}^{\mathrm{T}}(n)\}E\{\xi(n)\} \tag{7.4.7}$$

is simplified by incorporating the *independence assumption*, which states: the present observation of the data (x(n), $d(n)$) are independent of the past observations (x(n–1), $d(n$–1)), (x(n–2), $d(n$–2)), ... where

$$\mathbf{x}(n) = [x(n)\ x(n-1)\ x(n-2)\ \cdots\ x(n-N+1)] \tag{7.4.8}$$

$$\mathbf{x}(n-1) = [x(n-1)\ x(n-2)\ \cdots\ x(n-N)] \tag{7.4.9}$$

Another way to justify the independence assumption is through the following observation: The LMS coefficients $\mathbf{w}(n)$ at any given time are affected by the whole past history of the data (x(n – 1), $d(n$ – 1)), (x(n – 2), $d(n$ – 2)), ... and, therefore, for smaller step-size parameter μ, the past N observations of the data have small contribution to $\mathbf{w}(n)$ and, thus, we can say that $\mathbf{w}(n)$ and

$\mathbf{x}(n)$ are weakly dependent. This observation clearly suggests the approximation given by (7.4.7). Substituting (7.4.7) into (7.4.6) we obtain

$$\begin{aligned} E\{\xi(n+1)\} &= [\mathbf{I} - 2\mu E\{\mathbf{x}(n)\mathbf{x}^{\mathrm{T}}(n)\}]E\{\xi(n)\} \\ &= [\mathbf{I} - 2\mu \mathbf{R}_{\mathrm{x}}]E\{\xi(n)\} \end{aligned} \tag{7.4.10}$$

The mathematical forms of (7.4.10) and (6.2.5) of the steepest-descent method are identical except that the deterministic weight-error vector $\xi(n)$ in (6.2.5) has been replaced by the average weight-error vector $E\{\xi(n)\}$ of the LMS algorithm. This suggests that, on the average, the present LMS algorithm behaves just like the steepest descent algorithm. Like the steepest-descent method, the LMS algorithm is controlled by M modes of convergence, which are dependent on the eigenvalues of the correlation matrix $\mathbf{R}_{\mathrm{x}}$. In particular, the convergence behavior of the LMS algorithm is directly related to the eigenvalue spread of $\mathbf{R}_{\mathrm{x}}$ and, hence, to the power spectrum of the input data $\mathbf{x}(n)$. The more flatness of the power spectrum, the higher speed of convergence of the LMS algorithm is attained.

Learning Curve

In the development below we assume the following: a) the input signal to LMS filter $x(n)$ is zero-mean stationary process; b) the desired signal $d(n)$ is zero-mean stationary process; c) $\mathrm{x}(n)$ and $d(n)$ are jointly Gaussian-distributed random variables for all n; and d) at times n, the coefficients $\mathbf{w}(n)$ are independent of the input vector $\mathrm{x}(n)$ and the desired signal $d(n)$. The validity of d) is justified for small values of μ (independent assumption). Assumptions a) and b) simplify the analysis. Assumption c) simplifies the final results so that the third-order and higher moments that appear in the derivation are expressed in terms of the second-order moments due to their Gaussian distribution.

If we take the mean-square average of the error given by (7.4.3), we obtain

$$\begin{aligned} J(n) = E\{e^2(n)\} &= E\{[e^o(n) - \xi^T(n)\mathbf{x}(n)][e^o(n) - \mathbf{x}^{\mathrm{T}}(n)\xi(n)]\} \\ &= E\{e^{o2}(n)\} + E\{\xi^T(n)\mathbf{x}(n)\mathbf{x}^{\mathrm{T}}(n)\xi(n)\} - 2E\{e^o(n)\xi^T(n)\mathbf{x}(n)\} \end{aligned} \tag{7.4.11}$$

For independent random variables we have the following relations:

$$E\{xy\} = E\{x\}E\{y\} = E\{xE\{y\}\} \tag{7.4.12}$$

$$E\{x^2y^2\} = E\{x^2\}E\{y^2\} = E\{x^2E\{y^2\}\} = E\{xE\{y^2\}x\} \tag{7.4.13}$$

Based on the above two equations, the second term of (7.4.11) becomes ($\mathbf{x}^T(n)\xi(n) = \xi^T(n)\mathbf{x}(n) = number$)

$$
\begin{aligned}
E\{(\xi^T(n)\mathbf{x}(n))^2\} &= E\{\xi^T(n)\mathbf{x}(n)\mathbf{x}^T(n)\xi(n)\} = E\{\xi^T(n)E\{\mathbf{x}(n)\mathbf{x}^T(n)\}\xi(n)\} \\
&= E\{\xi^T\mathbf{R}_x\xi(n)\} = tr\{E\{\xi^T(n)\mathbf{R}_x\xi(n)\}\} \\
&= E\{tr\{\xi^T(n)\mathbf{R}_x\xi(n)\}\} = E\{tr\{\xi(n)\xi^T(n)\mathbf{R}_x\}\} \\
&= tr\{E\{\xi(n)\xi^T(n)\}\mathbf{R}_x\} = tr\{\mathbf{K}(n)\mathbf{R}_x\}
\end{aligned} \tag{7.4.14}
$$

where we used the following properties: a) the trace of a scalar is the scalar itself, b) the trace, *tr*, and the expectation, *E*, operators are linear and can be exchanged, c) the trace of two matrices having $M \times N$ and $N \times M$ dimensions, respectively, is given by

$$
tr\{\mathbf{AB}\} = tr\{\mathbf{BA}\} \tag{7.4.15}
$$

The third term in (7.4.11), due to the independence assumption and due to the fact that $e^o(n)$ is a constant, becomes

$$
E\{e^o(n)\xi^T(n)\mathbf{x}(n)\} = E\{\xi^T(n)\mathbf{x}(n)e^o(n)\} = E\{\xi^T(n)\}E\{\mathbf{x}(n)e^o(n)\} = 0 \tag{7.4.16}
$$

The second term is equal to zero due to the orthogonality property (see Section 4.3).

Substituting (7.4.16) and (7.4.15) in (7.4.11) we obtain

$$
J(n) = E\{e^2(n)\} = J_{\min} + tr\{\mathbf{K}(n)\mathbf{R}_x\} \tag{7.4.17}
$$

where $J_{\min} = E\{(e^o(n))^2\}$ is the minimum mean-square error. However, $\mathbf{R}_x = \mathbf{Q}\Lambda\mathbf{Q}^T$, where $\mathbf{Q}$ is the eigenvector matrix and Λ is the diagonal eigenvalue one (see Section 5.1). Hence, (7.4.17) becomes

$$
\begin{aligned}
J(n) &= J_{\min} + tr\{\mathbf{K}(n)\mathbf{Q}\Lambda\mathbf{Q}^T\} = J_{\min} + tr\{\mathbf{Q}^T\mathbf{K}(n)\mathbf{Q}\Lambda\} \\
&= J_{\min} + tr\{E\{\mathbf{Q}^T\xi(n)\xi^T(n)\mathbf{Q}\}\Lambda\} \\
&= J_{\min} + tr\{E\{\xi'(n)\xi'^T(n)\}\Lambda\} = J_{\min} + tr\{\mathbf{K}'(n)\Lambda\}
\end{aligned} \tag{7.4.18}
$$

where

$$
\mathbf{K}'(n) = E\{\xi'(n)\xi'^T(n)\} \tag{7.4.19}
$$

Recall that $\xi'(n)$ is the weight-error vector in the coordinate system defined by the basis vectors, which are specified by the eigenvectors of **R**. Since Λ is diagonal, (7.4.18) becomes

$$
J(n) = J_{\min} + \sum_{i=0}^{M-1} \lambda_i \kappa'_{ii}(n) = J_{\min} + \sum_{i=0}^{M-1} \lambda_i E\{\xi'^2_{ii}(n)\} \tag{7.4.20}
$$

where $\kappa'_{ij}(n)$ is the ij^{th} element of the matrix $\mathbf{K}'(n)$.

The learning curve can be obtained by any one of the equations: (7.4.17), (7.4.18), or (7.4.20). It turns out that, on the average, the learning curve above is similar to the one given by the steepest-descent algorithm.

The general solution of $\xi'(n)$ is given by (6.2.15). Hence, (7.4.20) becomes

$$J(n) = J_{\min} + \sum_{i=0}^{M-1} \lambda_i (1 - 2\mu\lambda_i)^{2n} E\{\xi'_{ii}(0)\} \tag{7.4.21}$$

Example 7.4.1: The filter $H_1(z)$, which produces the desired signal $d(n) = x(n)$, is represented by the difference equation $x(n) + a_1x(n-1) + a_2x(n-2) = v(n)$, where a_1 and a_2 are the system coefficients, and $v(n)$ is a zero-mean white noise process of variance σ_v^2. To obtain a_1 and a_2, we use the adaptive predictor. The LMS algorithm for this case is given by ($\mu' = 2\mu$)

$$\mathbf{w}(n) = \mathbf{w}(n) + \mu'\mathbf{x}(n-1)e(n) \tag{7.4.22}$$

where

$$e(n) = d(n) - \mathbf{w}^T(n)\mathbf{x}(n-1) = x(n) - \mathbf{w}^T(n)\mathbf{x}(n-1) \tag{7.4.23}$$

The learning curves, $J(n) = E\{e^2(n)\}$, are shown in Figure 7.4.1 using the following constants: M = 2, $a_1 = -0.96$, $a_2 = 0.2$, $\sigma_v^2 = 0.33$, $\mu' = 0.04$, and $\mu' = 0.004$. The curves were averaged over 120 runs. The following MATLAB function was used:

Book MATLAB function for Example 7.4.1

```
function[ems]=aaexample741(mu,M,an)
%function[ems]=aaexample741(mu,M);
%this function produces figures like in example 7.4.1;
```

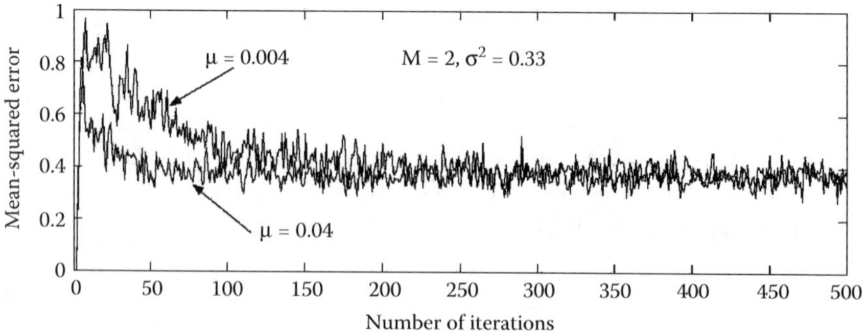

Figure 7.4.1 Learning curves for two different step-size parameter values.

```
%M=number of filter coefficients;an=number of times the
%error to be averaged;
eav=zeros(1,1000);
dn(1)=0;dn(2)=0;x(1)=0;x(2)=0;
for k=1:an
      for n=3:1000
            dn(n)=0.96*dn(n-1)-0.2*dn(n-2)+2*(rand-0.5);
            x(n)=dn(n-1);
      end;
      [w,y,e]=aalms(x,dn,mu,M);
      eav=eav+e.^2;
end;
ems=eav/an;
```

The coefficient-error or weighted-error correlation matrix

Since the mean-square error (MSE) is related to weight-error correlation matrix $\mathbf{K}(n)$, the matrix is closely related to the stability of the LMS algorithm. Therefore, $J(n)$ is bounded if the elements of $\mathbf{K}(n)$ are also bounded. Since $\mathbf{K}(n)$ and $\mathbf{K}'(n)$ are related, the stability can be studied by using either one.

Multiply both sides of (7.4.5) by $\mathbf{Q}^{\mathrm{T}}$ and use the definitions $\xi'(n) = \mathbf{Q}^{\mathrm{T}}\xi(n)$ and $\mathbf{x}'(n) = \mathbf{Q}^{\mathrm{T}}\mathbf{x}(n)$ to obtain (see Problem 7.4.1)

$$\xi'(n+1) = (\mathbf{I} - 2\mu\mathbf{x}'(n)\mathbf{x}'^{\mathrm{T}}(n))\xi'(n) + 2\mu e^{o}(n)\mathbf{x}'(n) \tag{7.4.24}$$

Next multiply (7.4.24) by its own transpose, and take the ensemble average of both sides to obtain (see Problem 7.4.2)

$$\begin{aligned}
\mathbf{K}'(n+1) = \mathbf{K}'(n) &- 2\mu E\{\mathbf{x}'(n)\mathbf{x}'^{\mathrm{T}}(n)\xi'(n)\xi'^{\mathrm{T}}(n)\} \\
&- 2\mu E\{\xi'(n)\xi'^{\mathrm{T}}(n)\mathbf{x}'(n)\mathbf{x}'^{\mathrm{T}}(n)\} \\
&+ 4\mu^2 E\{\mathbf{x}'(n)\mathbf{x}'^{\mathrm{T}}(n)\xi'(n)\xi'^{\mathrm{T}}(n)\mathbf{x}'(n)\mathbf{x}'^{\mathrm{T}}(n)\} \\
&+ 2\mu E\{e^{o}(n)\mathbf{x}'(n)\xi'^{\mathrm{T}}(n)\} \\
&+ 2\mu E\{e^{o}(n)\xi'(n)\mathbf{x}'^{\mathrm{T}}(n)\} \\
&- 4\mu^2 E\{e^{o}(n)\mathbf{x}'(n)\xi'^{\mathrm{T}}(n)\mathbf{x}'(n)\mathbf{x}'^{\mathrm{T}}(n)\} \\
&- 4\mu^2 E\{e^{o}(n)\mathbf{x}'(n)\mathbf{x}'^{\mathrm{T}}(n)\xi'(n)\mathbf{x}'^{\mathrm{T}}(n)\} \\
&+ 4\mu^2 E\{e^{o2}(n)\mathbf{x}'(n)\mathbf{x}'^{\mathrm{T}}(n)\}
\end{aligned} \tag{7.4.25}$$

Based on the previous independent assumption, we note the following: (1) (n) is independent of the data $\mathbf{x}(n)$ and the desired signal $d(n)$, which is also true for the transformed variables $\mathbf{x}'(n)$; 2) $\xi'(n)$ is independent of $\mathbf{x}'(n)$ and $d(n)$; 3) $d(n)$ and $\mathbf{x}'(n)$ are zero-mean and are jointly Gaussian since $d(n)$ and x(n) have the same properties; 4) applying the orthogonality relationship, we obtain $E\{e^o(n)\mathbf{x}'(n)\} = E\{e^o(n)\mathbf{Q}^T\mathbf{x}(n)\} = \mathbf{Q}^T E\{e^o(n)\mathbf{x}(n)\} = \mathbf{Q}^T\mathbf{0} = \mathbf{0}$;5) $e^o(n)$ depends only on $d(n)$ and $\mathbf{x}(n)$; 6) from 4) $e^o(n)$ and $\mathbf{x}'(n)$ are uncorrelated (Gaussian variables are also independent); 7) $e^o(n)$ has zero mean. With the above assumptions in mind, the factors of (7.4.25) become as follows:

$$\begin{aligned} E\{\mathbf{x}'(n)\mathbf{x}'^T(n)\xi'(n)\xi'^T(n)\} &= E\{\mathbf{x}'\mathbf{x}'^T(n)\}E\{\xi'(n)\xi'^T(n)\} \\ &= E\{\mathbf{Q}^T\mathbf{x}(n)\mathbf{x}^T(n)\mathbf{Q}\}\mathbf{K}'(n) = \mathbf{Q}^T\mathbf{R}_x\mathbf{Q}\mathbf{K}'(n) = \Lambda\mathbf{K}'(n) \end{aligned} \tag{7.4.26}$$

$$E\{\xi'(n)\xi'^T(n)\mathbf{x}'(n)\mathbf{x}'^T(n)\} = \mathbf{K}'(n)\Lambda \tag{7.4.27}$$

$$E\{\mathbf{x}'(n)\mathbf{x}'^T(n)\xi'(n)\xi'^T(n)\mathbf{x}'(n)\mathbf{x}'(n)\} = 2\Lambda\mathbf{K}'(n)\Lambda + tr\{\Lambda\mathbf{K}'(n)\Lambda \tag{7.4.28}$$

(see Problem 7.4.3).

Because $e^o(n)$, $\mathbf{x}'(n)$, and $\xi'(n)$ are mutually independent, and $E\{e^o(n)\} = 0$ then (7.4.29) to (7.4.32) are true.

$$E\{e^o(n)\mathbf{x}'(n)\xi'^T(n)\} = E\{e^o(n)\}E\{\mathbf{x}'(n)\xi'^T(n)\} = 0 \quad (M \times M \ matrix) \tag{7.4.29}$$

$$E\{e^o(n)\xi'(n)\mathbf{x}'^T(n)\} = 0 \tag{7.4.30}$$

$$E\{e^o(n)\mathbf{x}'(n)\xi'^T(n)\mathbf{x}'(n)\mathbf{x}'^T(n)\} = 0 \tag{7.4.31}$$

$$E\{e^o(n)\mathbf{x}'(n)\mathbf{x}'^T(n)\xi'(n)\mathbf{x}'^T(n)\} = 0 \tag{7.4.32}$$

$$E\{e^{o2}(n)\mathbf{x}'(n)\mathbf{x}'^T(n)\} = E\{e^{o2}\}E\{\mathbf{x}'(n)\mathbf{x}'^T(n)\} = J_{\min}\Lambda \tag{7.4.33}$$

Substituting (7.4.26)–(7.4.33) in (7.4.25) we find

$$\begin{aligned} \mathbf{K}'(n+1) = \mathbf{K}'(n) - 2\mu(\Lambda\mathbf{K}'(n) + \mathbf{K}'(n)\Lambda) + 8\mu^2\,\Lambda\mathbf{K}'(n)\Lambda \\ + 4\mu^2 tr\{\Lambda\mathbf{K}'(n)\}\Lambda + 4\mu^2 J_{\min}\,\Lambda \end{aligned} \tag{7.4.34}$$

Concentrating on the ii^{th} component of both sides of (7.4.34) we find (see Problem 7.4.4)

$$k'_{ii}(n+1) = (1 - 4\mu\lambda_i + 8\mu^2\lambda_i^2)k'_{ii}(n) + 4\mu^2\lambda_i \sum_{j=0}^{M-1} \lambda_j k'_{jj}(n) + 4\mu^2 J_{\min}\lambda_i \tag{7.4.35}$$

Since $\mathbf{K}'(n)$ is a correlation matrix, $k'^2_{ij} \le k'_{ii}(n)k'_{jj}(n)$ for all values of i and j, and since the update of $k'_{ii}(n)$ is independent of the off-diagonal elements of $K'(n)$, the convergence of the diagonal elements is sufficient to secure the convergence of all the elements and, thus, guarantees the stability of the LMS algorithm. Therefore, we concentrate on (7.4.35) with $i = 0, 1, 2, \ldots, M-1$.

Equation (7.4.35) can be written in the following matrix form (see Problem 7.4.4)

$$\mathbf{k}'(n+1) = \mathbf{F}\mathbf{k}'(n) + 4\mu^2 J_{\min} \boldsymbol{\lambda} \tag{7.4.36}$$

where

$$\mathbf{k}'(n) = [k'_{00}(n) \quad k'_{11}(n) \quad \cdots \quad k'_{M-1,M-1}(n)]^{\mathrm{T}} \tag{7.4.37}$$

$$\boldsymbol{\lambda} = [\lambda_0 \quad \lambda_1 \quad \cdots \quad \lambda_{M-1}]^{\mathrm{T}} \tag{7.4.38}$$

$$\mathbf{F} = diag[f_0 \; f_1 \quad \cdots \quad f_{M-1}] + 4\mu^2 \boldsymbol{\lambda}\, \boldsymbol{\lambda}^{\mathrm{T}} \tag{7.4.39}$$

$$f_i = 1 - 4\mu\lambda_i + 8\mu^2 \lambda_i^2 \tag{7.4.40}$$

It has been found that if the eigenvalues of **F** are all less than one, the LMS algorithm is stable or equivalently the elements $\mathbf{k}'(n)$ remain bounded. An indirect approach to obtain stability is given below.

Excess mean-square error and misadjustment

According to (7.4.17), we may write the expression of the difference between the mean-squared error and the minimum mean-square error as follows:

$$J_{ex}(n) = J(n) - J_{\min} = tr\{\mathbf{K}(n)\mathbf{R}_x\} \tag{7.4.41}$$

The steady state form of (7.4.41) is

$$J_{ex}(\infty) = J(\infty) - J_{\min} = tr\{\mathbf{K}(\infty)\mathbf{R}_x\} \tag{7.4.42}$$

and it is known as *excess MSE*. Equation (7.4.20) gives another equivalent form, which is

$$J_{ex}(\infty) = \sum_{i=0}^{M-1} \lambda_i k'_{ii}(\infty) = \xi^{\mathrm{T}} \mathbf{k}'(\infty) \tag{7.4.43}$$

As $n \to \infty$, we set $\mathbf{k}'(n+1) = \mathbf{k}'(n)$ and, hence, (7.4.36) becomes

$$\mathbf{k}'(\infty) = 4\mu^2 J_{\min} (\mathbf{I} - \mathbf{F})^{-1} \boldsymbol{\lambda} \tag{7.4.44}$$

As a result, (7.4.43) becomes

$$J_{ex}(\infty) = 4\mu^2 J_{\min} \boldsymbol{\lambda}^{\mathrm{T}} (\mathbf{I} - \mathbf{F})^{-1} \boldsymbol{\lambda} \tag{7.4.45}$$

which indicates that J_{ex} is proportional to $J_{\min}$. The normalized $J_{ex}(\infty)$ is equal to

$$\mathcal{M} = \frac{J_{ex}(\infty)}{J_{\min}} = 4\mu^2 \boldsymbol{\lambda}^{\mathrm{T}} (\mathbf{I} - \mathbf{F})^{-1} \boldsymbol{\lambda} \tag{7.4.46}$$

which is known as the *misadjustment* factor.

If $\mathbf{A}(N \times N)$, $\mathbf{B}(M \times M)$, and $\mathbf{C}(N \times M)$ are matrices that have inverses, then

$$(\mathbf{A} + \mathbf{C}\mathbf{B}\mathbf{C}^{\mathrm{T}})^{-1} = \mathbf{A}^{-1} - \mathbf{A}^{-1}\mathbf{C}(\mathbf{B}^{-1} + \mathbf{C}^{T}\mathbf{A}^{-1}\mathbf{C})^{-1}\mathbf{C}^{T}\mathbf{A}^{-1} \tag{7.4.47}$$

But

$$\mathbf{I} - \mathbf{F} = \begin{bmatrix} 1 - f_0 & 0 & \cdots 0 \\ 0 & 1 - f_1 & \cdots 0 \\ \vdots & & \\ 0 & 0 & \cdots 1 - f_{M-1} \end{bmatrix} = 4\mu^2 \boldsymbol{\lambda}\boldsymbol{\lambda}^T = \mathbf{F}_1 + a\boldsymbol{\lambda}\boldsymbol{\lambda}^T \tag{7.4.48}$$

where $a = -4\mu^2$. Therefore, (7.4.46) takes the form (see Problem 7.4.5)

$$\begin{aligned} \mathcal{M} &= -a\boldsymbol{\lambda}^{\mathrm{T}}(\mathbf{F}_1 + a\boldsymbol{\lambda}\boldsymbol{\lambda}^{\mathrm{T}})^{-1}\boldsymbol{\lambda} = -a\boldsymbol{\lambda}^{\mathrm{T}}\left[\mathbf{F}_1^{-1} - \frac{a\mathbf{F}_1^{-1}\boldsymbol{\lambda}\boldsymbol{\lambda}^{\mathrm{T}}\mathbf{F}_1^{-1}}{1 + a\boldsymbol{\lambda}^T\mathbf{F}_1^{-1}\boldsymbol{\lambda}}\right]\boldsymbol{\lambda} \\ &= a'\boldsymbol{\lambda}^{\mathrm{T}}\left[\mathbf{F}_1^{-1} + \frac{a'\mathbf{F}_1^{-1}\boldsymbol{\lambda}\boldsymbol{\lambda}^{\mathrm{T}}\mathbf{F}_1^{-1}}{1 - a'\boldsymbol{\lambda}^T\mathbf{F}_1^{-1}\boldsymbol{\lambda}}\right]\boldsymbol{\lambda} = \frac{\sum_{i=0}^{M-1} \frac{\mu\lambda_i}{1-2\mu\lambda_i}}{1 - \sum_{i=0}^{M-1} \frac{\mu\lambda_i}{1-2\mu\lambda_i}} \end{aligned} \tag{7.4.49}$$

where in (7.4.47) we set $\mathbf{C} = \boldsymbol{\lambda}$, $\mathbf{B} = \mathbf{I}$, $a' = -a$, and $\mathbf{A} = \mathbf{F}_1$. Small $\mathcal{M}$ implies that the summation on the numerator is small. If, in addition, $2\mu\lambda_i << 1$, we obtain the following result

$$\sum_{i=0}^{M-1} \frac{\mu\lambda_i}{1 - \mu\lambda_i} \cong \mu \sum_{i=0}^{M-1} \lambda_i = \mu tr\{\mathbf{R}_{\mathbf{x}}\} \tag{7.4.50}$$

Hence, (7.4.49) becomes

$$\mathcal{M} = \frac{\mu tr\{\mathbf{R}_{\mathbf{x}}\}}{1 - \mu tr\{\mathbf{R}_{\mathbf{x}}\}} \tag{7.4.51}$$

In addition, for $\mathcal{M} \ll 0.1$ the quantity $\mu\, tr\{\mathbf{R}_x\}$ is small and (7.4.51) becomes

$$\mathcal{M} = \mu\, tr\{\mathbf{R}_x\} \tag{7.4.52}$$

Since $r_x(0)$ is the mean-squared value of the input signal to an M-coefficient filter, we write

$$\mathcal{M} = \mu M r_x(0) = \mu M E\{x^2(n)\} = \mu M(\text{Power input}) \tag{7.4.53}$$

The above equation indicates that to keep the misadjustment factor small and at a specific desired value as the signal power changes, we must adjust the value of μ.

Stability

If we set

$$\mathrm{L} = \sum_{i=0}^{M-1} \frac{\mu \lambda_i}{1 - 2\mu \lambda_i} \tag{7.4.54}$$

then

$$\mathcal{M} = \frac{\mathrm{L}}{1 - \mathrm{L}} \tag{7.4.55}$$

We observe that L and $\mathcal{M}$ are increasing functions of μ and L, respectively (see Problem 7.4.6). Since L reaches 1 as $\mathcal{M}$ goes to infinity, this indicates that there is a value $\mu_{\max}$ that μ cannot surpass. To find the upper value of μ we must concentrate on an expression that can easily be measured in practice. Therefore, from (7.4.55) we must have

$$\sum_{i=0}^{M-1} \frac{\mu \lambda_i}{1 - 2\mu \lambda_i} \le 1 \tag{7.4.56}$$

The above equation indicates that the maximum value of μ must make equation (7.4.56) an equality. It can be shown that the following inequality holds.

$$\sum_{i=0}^{M-1} \frac{\mu \lambda_i}{1 - 2\mu \lambda_i} \le \frac{\mu \sum_{i=0}^{M-1} \lambda_i}{1 - 2\mu \sum_{i=0}^{M-1} \lambda_i} \tag{7.4.57}$$

Hence, if we solve the equality

$$\frac{\mu\sum_{i=0}^{M-1}\lambda_i}{1-2\mu\sum_{i=0}^{M-1}\lambda_i}=1 \tag{7.4.58}$$

for μ, then (7.4.56) is satisfied. The solution of the above equality is

$$\mu_{max}=\frac{1}{3\sum_{i=0}^{M-1}\lambda_i}=\frac{1}{3tr\{\mathbf{R}_x\}}=\frac{1}{3(Input\ Power)} \tag{7.4.59}$$

and, hence,

$$0<\mu<\frac{1}{3tr\{\mathbf{R}_x\}} \tag{7.4.60}$$

LMS and the steepest-descent method

The following similarities and differences exist between the two methods:

1. The steepest-descent method reaches the minimum MSE J_{min} as $n \to \infty$ and $\mathbf{w}(n) \to \mathbf{w}^o$.
2. The LMS method produces an error $J(\infty)$ that approaches J_{min} as $n \to \infty$ and remains larger than J_{min}.
3. The LMS method produces a $\mathbf{w}(n)$, as the iterations $n \to \infty$, that is close to the optimum $\mathbf{w}^o$.
4. The steepest-descent method has a well-defined learning curve consisting of a sum of decaying exponentials.
5. The LMS learning curve is a sum of noisy decaying exponentials, and the noise, in general, decreases the small values the step-size parameter μ takes.
6. In the steepest-descent method, the correlation matrix $\mathbf{R}_x$ of the data $\mathbf{x}(n)$ and the cross-correlation vector $\mathbf{p}_{dx}(n)$ are found using ensemble averaging operations from the realizations of the data $\mathbf{x}(n)$ and desired signal $d(n)$.
7. In the LMS filter, an ensemble of learning curves is found under identical filter parameters and then averaged point by point.

Example 7.4.2 (Channel Equalization): Figure 7.4.2 shows a base-band data transmission system equipped with an adaptive channel equalizer and

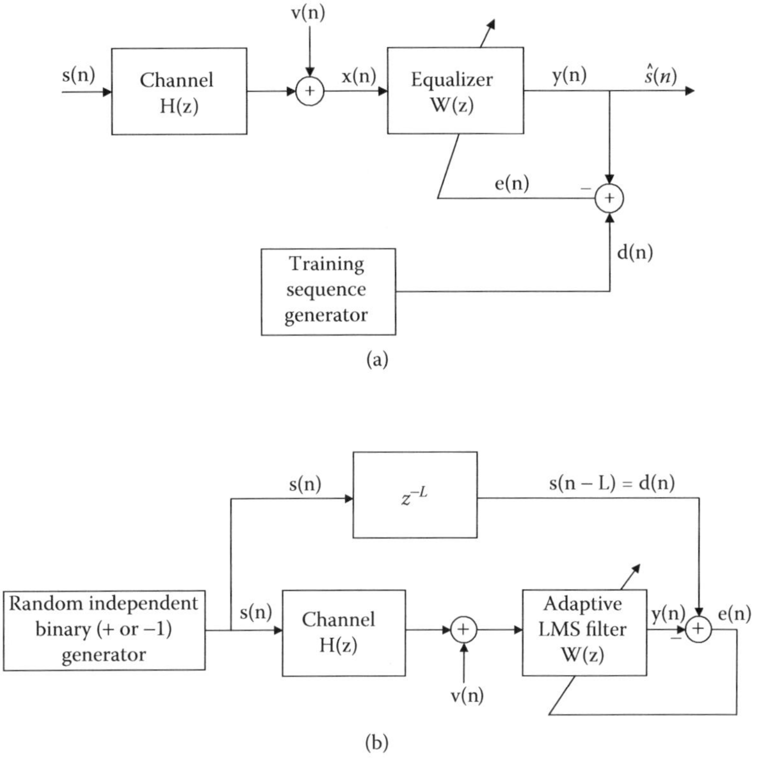

Figure 7.4.2 (a) Base-band data transmission system equipped with an adaptive channel equalizer and (b) a training system.

a training system. The signal $s(n)$ transmitted through the communication channel is amplitude or phase modulated pulses. The communication channel distorts the signal (the most important one is the pulse spreading) and results in overlapping of pulses and, thus, creating the *intersymbol interference* phenomenon. The noise $v(n)$ further deteriorates the fidelity of the signal. It is ideally required that the output of the equalizer is the signal $s(n)$. Therefore, an initialization period is used during which the transmitter sends a sequence of training symbols that are known to the receiver (*training mode*). This approach is satisfactory if the channel does not change characteristics rapidly in time. However, for slow changes the output from the channel can be treated as the desired signal for further adaptation of the equalizer so that its variations can be followed (*decision directed mode*).

If the equalization filter is the inverse of the channel filter, W(z) = 1/ H(z), the output will be that of the input to the channel, assuming, of course, that noise is small. To avoid singularities from the zero of the channel transfer function inside the unit circle, we select an equalizer such that $W(z)H(z) \cong z^{-L}$. This indicates that the output of the filter W(z) is that of the input to the channel shifted by L units of time. Sometimes, more general filters of the form Y(z)/H(z) are used where $Y(z) \neq z^{-L}$. These systems are known as the partial-response signaling systems. In these cases, Y(z) is selected such that the amplitude spectra are about equal over the range of frequencies of interest. The result of this choice is that W(z) has a magnitude response of about one, thereby minimizing the noise enhancement.

Figure 7.4.2b shows a channel equalization problem at training stage. The channel noise $v(n)$ is assumed to be white Gaussian with variance σ_v^2. The equalizer is an M-tap FIR filter and the desired output is assumed to be delayed replica of the signal $s(n)$, $s(n-L)$. The signal $s(n)$ is white, has variance $\sigma_s^2 = 1$, has zero mean value, and is uncorrelated with $v(n)$. The channel transfer function was assumed to take the following FIR forms:

$$\begin{aligned} H(z) &= H_1(z) = 0.34 + z^{-1} - 0.34z^{-2} \\ H(z) &= H_2(z) = 0.34 + 0.8z^{-1} + 0.1z^{-2} \end{aligned} \tag{7.4.61}$$

The solution of the two systems above, were selected based on the eigenvalue spread of their output correlation matrix. Figure 7.4.3a shows the learning curve with a channel system of the form $H(z) = 0.34 + z^{-1} - 0.34z^{-2}$

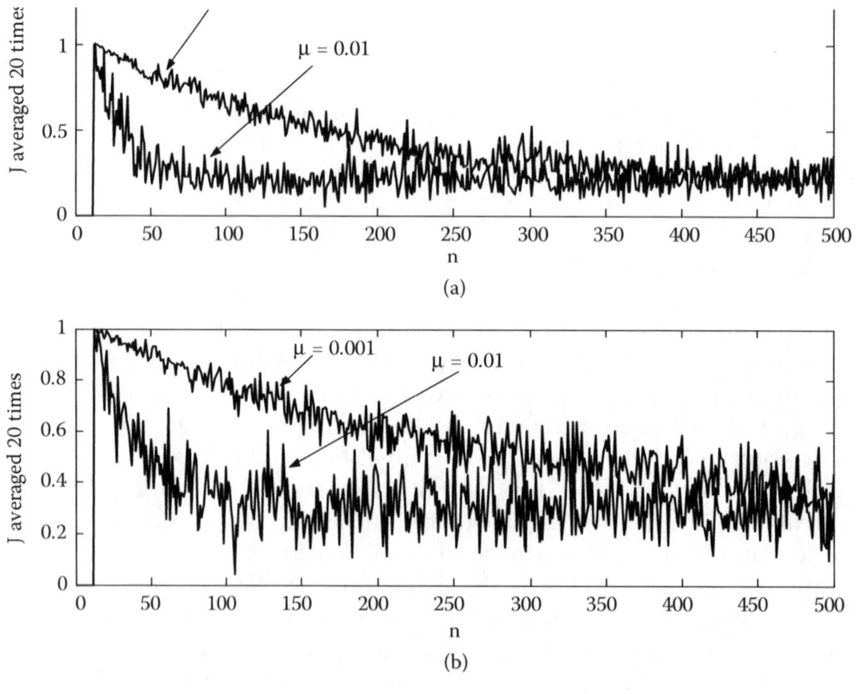

Figure 7.4.3 Learning curves of the channel system.

and two different step-size parameters and $c = 0.5$. The variance of the noise was $\sigma_v^2 = 0.043$. Figure 7.4.3b shows the learning curve with channel system of the form $H(z) = 0.34 + 0.8z^{-1} + 0.1z^{-2}$ and with the same step-size parameter and noise variance as in case (a) above. In both cases, the curves were reproduced 20 times and averaged. The delay L was assumed to be 3, and the number of filter coefficients were 12. The curves were produced using the following MATLAB function:

Book MATLAB function for channel equalization

```
function[Jav,wav,dn,e,x]=aaequalizer(av,M,L,h,N,mu,c)
%function[Jav,wav,dn,e,x]=aaequaliz-
%er(av,M,L,h,N,mu,c)
%this function solves the example depicted
%in Fig. 7.4.2;av=number of times to average e(error or
%learning curve)and w(filter coefficient);N=length
%of signal s;L=shift of the signal s to become
%dn;h=assumed
%impulse response of the channel system;mu=step fac-
%tor;
%M=number of adaptive filter coefficients;c=constant
%multiplier;
w1=[zeros(1,M)];
J=[zeros(1,N)];
for i=1:av
   for n=1:N
      v(n)=c*randn;
      s(n)=rand-.5;
      if s(n)<=0
         s(n)=-1;
      else
         s(n)=1;
      end;
   end;
      dn=[zeros(1,L)  s(:,1:N-L)];
      ych=filter(h,1,s);
      x=ych(1,1:N)+v;
      [w,y,e]=alms(x,dn,mu,M);
      w1=w1+w;
      J=J+e.^2;
   end;
   Jav=J/av;
   wav=w1/av;
```

It is recommended that the reader changes the input values to observe the effect on the learning curve.

7.5 Complex representation of LMS algorithm

In some practical applications it is mathematically attractable to have complex representation forms of the underlying signals. For example, base-band signals in quadrature amplitude modulation (QAM) format are written as a summation of two components: real *in-phase* component and imaginary *quadrature* component. Furthermore, signals detected by a set of antennas are also written in their complex form for easy mathematical manipulation. For this reason, we shall present in this section the most rudimentary derivation of the LMS filter assuming that the signals are complex.

In the case where complex-type signals must be processed, we write the output of the adaptive filter in the form

$$y(n) = \sum_{k=0}^{M-1} w_k^*(n)x(n-k) \tag{7.5.1}$$

and the error is given by

$$e(n) = d(n) - y(n) \tag{7.5.2}$$

Therefore, the MSE is

$$J = E\{e(n)e^*(n)\} = E\{|e(n)|^2\} \tag{7.5.3}$$

Let us define the complex filter coefficient as follows:

$$w_k = a_k(n) + jb_k(n) \qquad k = 0,1,\cdots,M-1 \tag{7.5.4}$$

Then the gradient operator ∇ has the following k^{th} element:

$$\nabla_{w_k} \triangleq \nabla_k = \frac{\partial}{\partial a_k(n)} + j\frac{\partial}{\partial b_k(n)} \qquad k = 0,1,\cdots,M-1 \tag{7.5.5}$$

which will produce the following k^{th} element of the multi-element gradient vector ∇J:

$$\nabla_k J = \frac{\partial J}{\partial a_k(n)} + j\frac{\partial J}{\partial b_k(n)} \tag{7.5.6}$$

It is noted that the gradient operator is always used to find the minimum points of a function. The above equation indicates that a complex constraint must be converted to a pair of real constraints. Hence, we set

$$\frac{\partial J}{\partial a_k(n)} = 0, \qquad j\frac{\partial J}{\partial b_k(n)} = 0 \tag{7.5.7}$$

The kth element of the gradient vector, using (7.5.3), is

$$\nabla_k J = E\{\frac{\partial e(n)}{\partial a_k(n)} e^*(n) + \frac{\partial e^*(n)}{\partial a_k(n)} e(n) + j\frac{\partial e(n)}{\partial b_k(n)} e^*(n) + j\frac{\partial e^*(n)}{\partial b_k(n)} e(n)\} \tag{7.5.8}$$

Taking into consideration (7.5.1) and (7.5.2), we obtain

$$\frac{\partial e(n)}{\partial a_k(n)} = \frac{\partial d(n)}{\partial a_k(n)} - \sum_{k=0}^{M-1} \frac{\partial w_k^*(n)}{\partial a_k(n)} x(n-k) = 0 - \sum_{k=0}^{M-1} \frac{\partial [a_k(0 - jb_k(n)]}{\partial a_k(n)} x(n-k) \tag{7.5.9}$$

which can be written as

$$\frac{\partial e(n)}{\partial a_k(n)} = -x(n-k) \tag{7.5.10}$$

$$\frac{\partial e^*(n)}{\partial b_k(n)} = jx(n-k), \quad \frac{\partial e^*(n)}{\partial a_k(n)} = -x^*(n-k), \quad \frac{\partial e^*(n)}{\partial b_k(n)} = -jx^*(n-k) \tag{7.5.11}$$

Introducing (7.5.10) and (7.5.11) into (7.5.8), we find the relationship

$$\nabla_k J \equiv \nabla_{w_k} J(\mathbf{w}(n)) = -2E\{x(n-k)e^*(n)\} \tag{7.5.12}$$

and, thus, the gradient vector becomes

$$\nabla_w J(\mathbf{w}(n)) = -2E\{e^*(n)[x(n)\ x(n-1)\ \cdots\ x(n-M+1)]^T\} \tag{7.5.13}$$

If $\mathbf{w}(n)$ is the filter-coefficient vector at step n (time), then its update value $\mathbf{w}(n+1)$ is given by (see (6.1.3))

$$\mathbf{w}(n+1) = \mathbf{w}(n) - \mu \nabla_w J(\mathbf{w}(n)) \tag{7.5.14}$$

Next, we replace the ensemble average in (7.5.13) by the instantaneous estimates $e^*(n)\mathbf{x}(n)$ to obtain

$$\mathbf{w}(n+1) = \mathbf{w}(n) + 2\mu e^*(n)\mathbf{x}(n) \tag{7.5.15}$$

Table 7.5.1 Complex Form of the LMS Algorithm

Parameters:	M = number of filter coefficients μ = step − size factor $\mathbf{x}(n) = [x(n)\ x(n-1) \cdots x(n-M+1)^T]$
Initialization:	$\mathbf{w} = \mathbf{0}$
Computation:	For $n = 0, 1, 2, \ldots$ 1. $y(n) = \mathbf{w}^H(n)\mathbf{x}(n)$ H: Hermitian = conjugate transpose 2. $e(\mathrm{n}) = d(n) - y(n)$ 3. $\mathbf{w}(n+1) = \mathbf{w}(n) + 2\mu e^*(n)\mathbf{x}(n)$

which is the LMS recursion formula when we are involved with complex-valued processes. For complex signal it has been shown that the misadjustment factor $\mathcal{M}$ is given by

$$\mathcal{M} = \frac{\sum_{k=0}^{M-1} \frac{\mu\lambda_k}{1-\mu\lambda_k}}{1 - \sum_{k=0}^{M-1} \frac{\mu\lambda_k}{1-\mu\lambda_k}} \tag{7.5.16}$$

provided, also, that

$$0 < \mu < \frac{1}{\lambda_{\max}} \tag{7.5.17}$$

and

$$\sum_{k=0}^{M-1} \frac{\mu\lambda_k}{1-\mu\lambda_k} < 1 \tag{7.5.18}$$

The LMS algorithm for complex signals is shown in Table 7.5.1.

Book MATLAB function for complex LMS algorithm

```
function[w,y,e,J,w1]=aacomplexnlms(x,dn,mubar,M,c)
%function[w,y,e,J,w1]=aacomplexnlms(x,dn,mubar,M,c)
%x=input data to the filter;dn=desired signal;
%M=filter order;c=constant;mubar=step-size equivalent
%parameter;
%x and dn must be of the same length;J=learning curve;
N=length(x);
y=zeros(1,N);
w=zeros(1,M);%initialized filter coefficient vector;
```

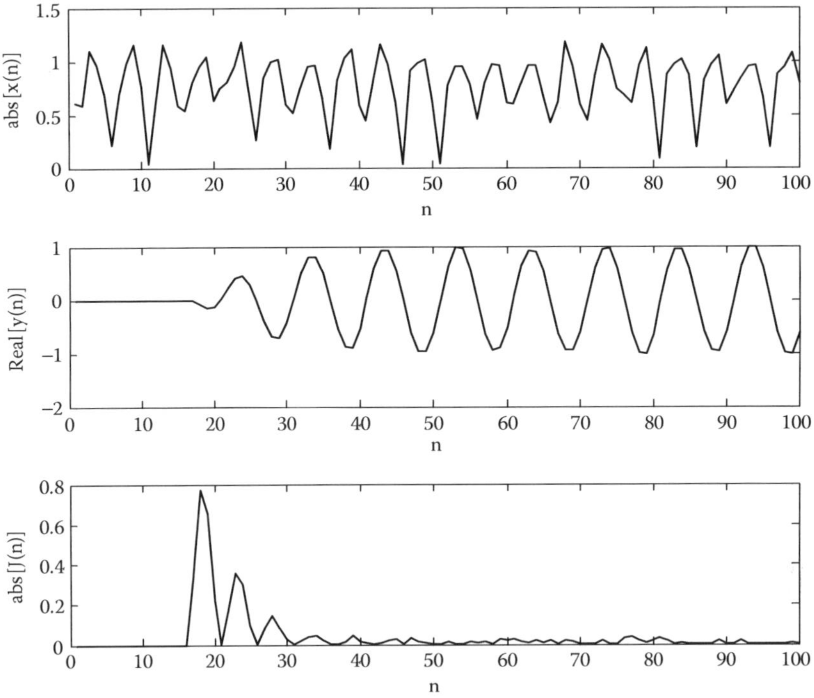

Figure 7.5.1 Results of Example 7.5.1 using the LMS complex-valued form.

```
for n=M:N
    x1=x(n:-1:n-M+1);%for each n vector x1 is of length
    %M with elements from x in reverse order;
    y(n)=conj(w)*x1';
    e(n)=dn(n)-y(n);
    w=w+(mubar/(c+conj(x1)*x1'))*conj(e(n))*x1;
    w1(n-M+1,:)=w(1,:);
end;
J=e.^2;
%the columns of the matrix w1 depict the history of the
%filter coefficients;
```

Example 7.5.1: With the input signal $x(n) = sin(0.2\pi n) + j1.5(\text{rand}-0.5)$, the desired signal $d(n) = sin(0.2\pi n)$, $\mu = 0.01$ and number of coefficients $M = 16$, we obtain the results shown in Figure 7.5.1.

Problems

7.2.1 Develop the LMS algorithm for complex-valued functions.

7.3.1 If an AR system has poles at $0.85e^{\pm j\pi/4}$ with input of white noise $v(n)$, find its discrete-time representation.

7.4.1 Verify (7.4.24).

7.4.2 Verify (7.4.25).

7.4.3 Verify (7.4.28).

7.4.4 Verify (7.4.36).

7.4.5 Verify (7.4.49).

7.4.6 Verify that (7.4.54) and (7.4.55) are increasing functions of μ and $\mathcal{L}$, respectively.

7.4.7 Use the *average eigenvalue* to obtain the *average time constant* for the LMS algorithms and state your observations for $\mathcal{M}$.

7.4.8 Discuss the effect of the initial value $\mathbf{w}(0)$ on the transient behavior of the LMS algorithm.

7.4.9 Develop the LMS algorithm using the steepest-descent method and setting $J(n) = e^2(n)$ instead of $E\{e^2(n)\}$.

7.4.10 (a) Let the impulse response of a FIR channel (see Figure 7.4.2b) have the values $\mathbf{h} = [0.22\ 1\ 0.22]^T$ and $n = 1, 2, 3$ and zero otherwise. The random binary generator produces a Bernoulli sequence with $s(n) = \pm 1$ having zero mean a unit variance. The sequence $v(n)$ is white with zero mean and variance $\sigma_v^2 = 0.01$. The equalizer is FIR filter with 5 coefficients. Find the correlation matrix $\mathbf{R}_x$ and the eigenvalues. Next find the step-size parameter based on the spread of the eigenvalues. b) Repeat part a) but with $\mathbf{h} = [0.39\ 1\ 0.39]^T$.

Hints-solutions-suggestions

7.2.1:

$$\mathbf{R}_{ix} = \mathbf{x}(n)\mathbf{x}^H(n),\ \mathbf{p}_{idx} = \mathbf{x}(n)d^*(n),\ \nabla J = -2\mathbf{x}(n)d^*(n) + 2\mathbf{x}(n)\mathbf{x}^H(n)\mathbf{w}(n)\ ,$$

$$\mathbf{w}(n+1) = \mathbf{w}(n) + 2\mu\mathbf{x}(n)[d^*(n) - \mathbf{x}^H(n)\mathbf{w}(n)],\ y(n) = \mathbf{w}^H(n)\mathbf{x}(n) = \mathbf{x}^H(n)\mathbf{w}(n)$$

$e(n) = d(n) - y(n)$, $\mathbf{w}(n+1) = \mathbf{w}(n) + 2\mu\mathbf{x}(n)e^*(n)$, H = conjugate transpose (or Hermitian).

7.3.1:

$H(z) = X(z)/V(z) = 1/[(1-0.85e^{j\pi/4}z^{-1})(1-0.85e^{-j\pi/4}z^{-1})]$ or $V(z) = X(z)[1 - 1.7\frac{\sqrt{2}}{2}z^{-1} + (0.85)^2 z^{-2}]$. Hence, the inverse Z-transform gives $v(n) = x(n) - 1.202x(n-1) + 0.7225x(n-2)$.

7.4.1:

$$\mathbf{Q}^{\mathbf{T}}\xi(n+1) = \mathbf{Q}^{\mathbf{T}}[\xi(n) - 2\mu\mathbf{x}(n)\mathbf{x}^{\mathbf{T}}(n)\xi(n)] + 2\mu e^{o}(n)\mathbf{Q}^{\mathbf{T}}\mathbf{x}(n) = \mathbf{Q}^{\mathbf{T}}\xi(n)$$

$$-2\mu\mathbf{Q}^{\mathbf{T}}\mathbf{x}(n)\mathbf{x}^{\mathbf{T}}(n)\mathbf{Q}^{-\mathbf{T}}\mathbf{Q}^{\mathbf{T}}\xi(n)] + 2\mu e^{o}(n)\mathbf{x}'(n) = \xi'(n) - 2\mu\mathbf{Q}^{\mathbf{T}}\mathbf{x}(n)(\mathbf{Q}^{\mathbf{T}}\mathbf{x}(n))^{T}\xi'(n)$$

$$+2\mu e^{o}(n)\mathbf{x}'(n) = [\mathbf{I} - 2\mu\mathbf{x}'(n)\mathbf{x}'^{\mathbf{T}}(n)]\xi'(n) + 2\mu e^{o}(n)\mathbf{x}'(n).$$

Note: $\mathbf{Q}^{-\mathbf{T}} = (\mathbf{Q}^{-1})^{T} = (\mathbf{Q}^{\mathbf{T}})^{T} = \mathbf{Q}.$

7.4.2:

Hint: $\xi'^{T}(n+1) = \xi'^{T}(n) - 2\mu\,\xi'^{T}(n)\mathbf{x}'(n)\mathbf{x}'^{\mathbf{T}}(n) + 2\mu e^{o}(n)\mathbf{x}'^{\mathbf{T}}(n).$

7.4.3:

Because $\mathbf{x}'(n)$ and $\xi'(n)$ are independent $E\{\mathbf{x}'(n)\mathbf{x}'^{\mathbf{T}}(n)\xi'(n)\xi'^{T}\mathbf{x}'(n)\mathbf{x}'^{\mathbf{T}}(n)\}$

$$= E\{\mathbf{x}'(n)\mathbf{x}'^{\mathbf{T}}(n)E\{\xi'(n)\xi'^{T}(n)\}\mathbf{x}'(n)\mathbf{x}'^{\mathbf{T}}(n)\} = E\{\mathbf{x}'(n)\mathbf{x}'^{\mathbf{T}}(n)\mathbf{K}'(n)\mathbf{x}'(n)\mathbf{x}'^{\mathbf{T}}(n)\} \quad (1)$$

$$\mathbf{x}'^{\mathbf{T}}(n)\mathbf{K}'(n)\mathbf{x}'(n) = \sum_{i=0}^{M-1}\sum_{j=0}^{M-1} x_i'(n)x_j'(n)k_{ij}'(n) \quad (2)$$

$$\mathbf{C}(n) = M \times M \text{ matrix} = \mathbf{x}'(n)\mathbf{x}'^{\mathbf{T}}(n)\mathbf{K}'(n)\mathbf{x}'(n)\mathbf{x}'^{\mathbf{T}}(n) \quad (3)$$

$$c_{lm}(n) = x_l'(n)x_m'(m)\sum_{i=0}^{M-1}\sum_{j=0}^{M-1} x_i'(n)x_j'(n)k_{ij}'(n) \quad (4)$$

$$E\{c_{lm}(n)\} = \sum_{i=0}^{M-1}\sum_{j=0}^{M-1} E\{x_l'(n)x_m'(n)x_i'(n)x_j'(n)\}k_{ij}'(n) \quad (5)$$

For Gaussian random variables we have

$$E\{x_1x_2x_3x_4\} = E\{x_1x_2\}E\{x_3x_4\} + E\{x_1x_3\}E\{x_2x_4\} + E\{x_1x_4\}E\{x_2x_3\} \quad (6)$$

$$E\{x_i'(n)x_j'(n)\} = \lambda_i\delta(i-j) \quad (7)$$

The above equation is due to the relation $E\{\mathbf{Q}^T\mathbf{x}(n)\mathbf{x}^T(n)\mathbf{Q}\} = \mathbf{Q}^T\mathbf{R}\mathbf{Q} = \Lambda$, $\mathbf{R}\mathbf{q}_i = \lambda_i\mathbf{q}$

$$E\{x_l'(n)x_m'(n)x_i'(n)x_j'(n)\} = \lambda_l\lambda_i\delta(l-m)\delta(i-j) + \lambda_l\lambda_m\delta(l-i)\delta(m-j) + \lambda_l\lambda_m\delta(l-j)\,\delta(m-i) \quad (8)$$

Substitute (8) in (5) to find

$$E\{c_{lm}(n)\} = \sum_{i=0}^{M-1}\sum_{j=0}^{M-1}\lambda_l\lambda_i\delta(l-m)\,\delta(i-j)k_{ij}'(n) + \sum_{i=0}^{M-1}\sum_{j=0}^{M-1}\lambda_l\,\lambda_m\;\delta(l-i)\,\delta(m-j)k_{ij}'(n)$$

$$+\sum_{i=0}^{M-1}\sum_{j=0}^{M-1}\lambda_l\,\lambda_m\;\delta(l-j)\,\delta(m-i)k_{ij}'(n) = \lambda_l\;\delta(l-m)\sum_{i=0}^{M-1}\lambda_i k_{ii}'(n) + \lambda_l\,\lambda_m k_{lm}'(n)$$

$+\lambda_l\,\lambda_m k_{ml}'(n)$ for $l = 0, 1, \ldots, M-1$ and $m = 0, 1, \ldots, M-1$. But $k_{lm}'(n) = k_{ml}'(n)$ and

$\sum_{i=0}^{M-1}\lambda_i k_{ii}'(n) = tr\{\mathbf{\Lambda K}'(n)\}$ and $\lambda_l\,\lambda_m k_{lm}'(n) + \lambda_l\,\lambda_m k_{ml}'(n) = 2\lambda_l\,\lambda_m k_{lm}'(n)$. Based on these results, (7.4.28) is apparent.

7.4.4:

$$\begin{bmatrix} k_{00}'(n+1) & k_{01}'(n+1)\cdots & \\ k_{10}'(n+1) & k_{11}'(n+1)\cdots & \\ \vdots & \vdots & \\ & & k_{M-1,M-1}'(n+1) \end{bmatrix} = \begin{bmatrix} f_0 + 4\mu^2\lambda_0^2 & 4\mu^2\lambda_1\lambda_0 & \cdots 4\mu^2\lambda_{M-1}\lambda_0 \\ \lambda_0\lambda_1 & f_1 + 4\mu^2\lambda_1^2 & \cdots 4\mu^2\lambda_{M-1}\lambda_1 \\ \vdots & \vdots & \vdots \\ \lambda_0\lambda_{M-1} & \lambda_1\lambda_{M-1} & \cdots f_{M-1} + 4\mu^2\lambda_{M-1}^2 \end{bmatrix}$$

$$\bullet \begin{bmatrix} k_{00}'(n+1) & k_{01}'(n+1) & \cdots & \\ k_{10}'(n+1) & k_{11}'(n+1) & \cdots & \\ \vdots & & & k_{M-1,M-1}'(n+1) \end{bmatrix} + 4\mu^2 J_{\min}\begin{bmatrix} \lambda_0 & & & \\ & \lambda_1 & 0 & \\ & 0 & \ddots & \\ & & & \lambda_{M-1} \end{bmatrix}$$

where the ii^{th} component of both sides is

$$k_{ii}'(n+1) = f_i k_{ii}'(n) + 4\mu^2\lambda_i[\lambda_0 k_{00}'(n) + 4\mu^2\lambda_1 k_{11}'(n) + \cdots + 4\mu^2\lambda_{M-1}k_{M-1}'(n)] + 4\mu^2 J_{\min}\lambda_i$$

which is identical to (7.4.36).

7.4.5:

The a' in (7.4.49) has been substituted with a in the development below.

$$a[\lambda_0\ \lambda_1]\begin{bmatrix}\frac{1}{1-f_0} & 0\\ 0 & \frac{1}{1-f_1}\end{bmatrix}\begin{bmatrix}\lambda_0\\ \lambda_1\end{bmatrix}+\frac{a^2[\lambda_0\ \lambda_1]\begin{bmatrix}\frac{1}{1-f_0} & 0\\ 0 & \frac{1}{1-f_1}\end{bmatrix}\begin{bmatrix}\lambda_0\lambda_0 & \lambda_0\lambda_1\\ \lambda_0\lambda_1 & \lambda_1\lambda_1\end{bmatrix}\begin{bmatrix}\frac{1}{1-f_1} & 0\\ 0 & \frac{1}{1-f_1}\end{bmatrix}\begin{bmatrix}\lambda_0\\ \lambda_1\end{bmatrix}}{1-a[\lambda_0\ \lambda_1]\begin{bmatrix}\frac{1}{1-f_0} & 0\\ 0 & \frac{1}{1-f_1}\end{bmatrix}\begin{bmatrix}\lambda_0\\ \lambda_1\end{bmatrix}}$$

$$=\frac{a\left[\frac{\lambda_0^2}{1-f_0}+\frac{\lambda_1^2}{1-f_1}\right]\left[1-a\left(\frac{\lambda_0^2}{1-f_0}+\frac{\lambda_1^2}{1-f_1}\right)\right]+()}{1-a\left(\frac{\lambda_0^2}{1-f_0}+\frac{\lambda_1^2}{1-f_1}\right)}$$

$$=\left[a\frac{\lambda_0^2}{1-f_0}+\frac{\lambda_1^2}{1-f_1}-a^2\frac{\lambda_0^4}{(1-f_0)^2}-a^2\frac{\lambda_1^4}{(1-f_1)^2}-2a^2\frac{\lambda_0^2\lambda_1^2}{(1-f_0)(1-f_1)}-2a^2\frac{\lambda_0^2}{(1-f_0)^2}\right.$$

$$\left.+a^2\frac{\lambda_0^2\lambda_1^2}{(1-f_0(1-f_1)}+a^2\frac{\lambda_0^2\lambda_1^2}{(1-f_0)(1-f_1)}+a^2\frac{\lambda_1^4}{(1-f_1)^2}\right]\Big/\left[1-a\left(\frac{\lambda_0^2}{1-f_0}+\frac{\lambda_1^2}{1-f_1}\right)\right]$$

$$=\frac{a\frac{\lambda_0^2}{1-f_0}+\frac{\lambda_1^2}{1-f_1}}{1-4\mu^2\frac{\lambda_0^2}{1-\left(1-4\mu\lambda_0+8\mu^2\lambda_0^2\right)}-4\mu^2\frac{\lambda_1^2}{1-\left(1-4\mu\lambda_1+8\mu^2\lambda_1^2\right)}}=\frac{\sum_{i=0}^{2-1}\frac{\mu\lambda_i}{1-2\mu\lambda_i}}{1-\sum_{i=0}^{2-1}\frac{\mu\lambda_i}{1-2\mu\lambda_i}}$$

which confirms (7.4.49) for $M = 2$.

7.4.6:

Hint: Take derivatives with respect to μ and L, respectively, and note the positive values of $\frac{\partial L}{\partial \mu}$ and $\frac{\partial \mathcal{M}}{\partial \mu}$.

7.4.7:

$$\lambda_{av}=\frac{1}{M}\sum_{i=0}^{M-1}\lambda_i=\frac{1}{M}tr\{\mathbf{R}_\mathbf{x}\},\ \tau_{mse,av}\cong\frac{1}{2\mu\lambda_{av}}\ \text{(see (6.2.12)) and,}$$

therefore, $\mathcal{M}\cong\mu M\lambda_{av}\cong M/(2\tau_{mse,av})$. Observations: (1) $\mathcal{M}$ increases linearly with filter length for a fixed λ_{av}, (2) the *setting time* of the LMS algorithm is proportional to $\tau_{mse,av}$ and, hence, $\mathcal{M}$ is inversely proportional to the settling time, (3) $\mathcal{M}$ is proportional to μ and inversely proportional to $\tau_{mse,av}$, but μ and $\tau_{mse,av}$ are inversely proportional to each other. Careful consideration must be given when we are considering values for $\mathcal{M}$, μ and $\tau_{mse,av}$.

7.4.8:

The averaged learning curve of the LMS algorithm is close to the steepest-descent method. Hence, we write (6.2.15) in the form:

$$J(n) \cong J_{\min} + \sum_{k=0}^{M-1} \lambda_k (1-2\mu\lambda_k)^{2n} \xi_k'^2(0) \tag{1}$$

If we set

$$\xi(0) = \mathbf{w}(n) - \mathbf{w}^o = 0 - \mathbf{w}^o, \text{ then } \xi'(n) = \mathbf{Q}^T \xi(0) = -\mathbf{Q}^T \mathbf{w}^o = -\mathbf{w}^o \tag{2}$$

Hence, (1) becomes

$$J(n) \cong J_{\min} + \sum_{k=0}^{M-1} \lambda_k (1-\mu\lambda_k)^{2n} w_k^{o\ \prime 2} \tag{3}$$

Equation (3) indicates that if $w_k^{o'}$'s corresponding to the smaller eigenvalues of $\mathbf{R}_x$ are all close to zero, then the transient behavior of the LMS algorithm is determined by the larger eigenvalues whose associated time constants are small, and thus, a fast convergence takes place. If, however, the $w_k^{o'}$'s corresponding to the smaller eigenvalues of the correlation matrix are significantly large, then we will observe that the slower modes will dominate.

7.4.9:

From (6.2.1) we find

$$\mathbf{w}(n+1) = \mathbf{w}(n) - \mu \nabla e^2(n) \tag{1}$$

where n is the iteration number, and

$$\nabla = \left[\frac{\partial}{\partial w_0} \quad \frac{\partial}{\partial w_1} \cdots \frac{\partial}{\partial w_{M-1}} \right]^T, \quad \mathbf{w}(n) = [w_0(n) \ w_1(n) \ \cdots \ w_{M-1}(n)]^T.$$

But

$$\partial e^2(n)/\partial w_i = 2e(n)\partial e(n)/\partial w_i = 2e(n)\partial[d(n) - y(n)]/\partial w_i = -2e(n)\partial y(n)/\partial w_i = -2e(n)$$

$$\partial\left[\sum_{i=0}^{M-1} w_i(n)x(n-i)\right]/\partial w_i(n) = -2e(n)x(n-i) \text{ and hence,}$$

$$\nabla e^2(n) = -2e(n)\mathbf{x}(n)$$

where

$$\mathbf{x}(n) = [x(n) \ x(n-1) \ x(n-2) \cdots x(n-M+1)]^T.$$

Therefore, (1) becomes

$$\mathbf{w}(n+1) = \mathbf{w}(n) + 2\mu e(n)\mathbf{x}(n)$$

7.4.10:

$$x(n)=\sum_{k=1}^{3}h(k)s(n-k)+v(n) \text{ and hence, we obtain}$$

$$E\{x(n)x(n-m)\}=r_x(m)=$$

$$E\{[\sum_{k=1}^{3}h(k)s(n-k)+v(n)][\sum_{k=1}^{3}h(i)s(n-m-i)+v(n-m)]\}$$

$$=\sum_{k}\sum_{i}h(k)h(i)E\{s(n-k)$$

$$s(n-m-i)\}+\sum_{i}h(i)E\{v(n)s(n-m-i)]\}$$

$$+\sum_{k}h(k)E\{s(n-k)v(n-m)\}+E\{v(n)v(n-m)\}$$

$$=\sum_{k}h(k)\sum_{i}h(i)\sigma_s^2\delta(m+i-k)+\sigma_v^2\delta(m)=\sigma_s^2\sum_{k=1}^{3}h(k)h(k-m)+\sigma_v^2\delta(m)$$

$$=\sigma_s^2 r_h(m)+\sigma_v^2\delta(m)$$

since $s(n)$ and $v(n)$ are uncorrelated with zero mean value. Hence,

$$r_x(0)=r_h(0)+\sigma_v^2=\sum_{k=1}^{3}h(k)h(k)+\sigma_v^2=h^2(1)+h^2(2)+h^2(3)+\sigma_v^2=1.168$$

$$r_x(1)=r_h(1)=\sum_{k=1}^{3}h(k)h(k-1)=h(2)h(1)+h(3)h(2)=0.44,\ r_x(2)=r_h(2)=0.0484$$

$r_x(3)=r_h(3)=0,\ r_x(4)=r_h(4)=0$ and the correlation matrix becomes

$$\mathbf{R}_X=\begin{bmatrix}1.168 & 0.44 & 0.0484 & 0 & 0\\ 0.44 & 1.168 & 0.44 & 0.0484 & 0\\ 0.0484 & 0.44 & 1.168 & 0.44 & 0.0484\\ 0 & 0.0484 & 0.44 & 1.168 & 0.44\\ 0 & 0 & 0.0484 & 0.44 & 1.168\end{bmatrix}.$$

The maximum and minimum eigenvalues are: λ_{min} = 0.7031, λ_{max} = 1.9869 and, hence, the eigenvalue spread is $\mathcal{X}(\mathbf{R}_x)=\lambda_{max}/\lambda_{min}$ = 4.3014. The value of the step-size parameter is found using (7.4.58):

$\mu=1/(3tr\{\mathbf{R}_x\})=0.095$. Part *b*) is found using similar steps.

chapter 8

Variations of LMS algorithms

8.1 The sign algorithms

This chapter covers the most popular modified LMS-type algorithms proposed by researchers over the past years, as well as some recent ones proposed by the authors. Most of these algorithms were designed on an ad hoc basis to improve convergence behavior, reduce computational requirements, and decrease the steady-state mean-square error. We start with this section by introducing the sign algorithms.

The error sign algorithm

The error sign algorithm is defined by

$$\mathbf{w}(n+1) = \mathbf{w}(n) + 2\mu \; sign(e(n))\mathbf{x}(n) \tag{8.1.1}$$

where

$$sign(n) = \begin{cases} 1 & n > 0 \\ 0 & n = 0 \\ -1 & n < 0 \end{cases} \tag{8.1.2}$$

is the *signum* function. By introducing the *signum* function and setting μ to a value of power of two, the hardware implementation is highly simplified (shift and add/subtract operation only).

Book MATLAB function for sign algorithm

```
function[w,y,e,J,w1]=aalmssign(x,dn,mu,M)
%function[w,y,e,J,w1]=aalmssign(x,dn,mu,M);
%all quantities are real-valued;
%x=input data to the filter;dn=desired signal;
%M=order of the filter;
```

```
%mu=step-size parameter;x and dn must be of the same length
N=length(x);
y=zeros(1,N);
w=zeros(1,M);%initialized filter coefficient vector
for n=M:N
   x1=x(n:-1:n-M+1);%for each n the vector x1 is produced
    %of length M with elements from x in reverse order;
    y(n)=w*x1';
    e(n)=dn(n)-y(n);
    w=w+2*mu*sign(e(n))*x1;
    w1(n-M+1,:)=w(1,:);
end;
J=e.^2;
%the columns of w1 depict the history of the filter
%coefficients;
```

The normalized LMS sign algorithm

The normalized LMS sign algorithm is (see below for normalized LMS algorithm)

$$\mathbf{w}(n+1) = \mathbf{w}(n) + 2\mu \frac{sign(e(n))\mathbf{x}(n)}{\varepsilon + \left\|\mathbf{x}(n)\right\|^2} \tag{8.1.3}$$

Book MATLAB function for normalized LMS sign algorithm

```
function[w,y,e,J,w1]=aanormlmssign(x,dn,mu,M,ep)
%function[w,y,e,J,w1]=aalmssign(x,dn,mu,M,ep);
%all quantities are real-valued;
%x=input data to the filter;dn=desired signal;
%M=order of the filter;
%mu=step-size parameter;x and dn must be of the same length;
%ep=sm
N=length(x);
y=zeros(1,N);
w=zeros(1,M);%initialized filter coefficient vector
for n=M:N
   x1=x(n:-1:n-M+1);%for each n the vector x1 is produced
    %of length M with elements from x in reverse order;
    y(n)=w*x1';
    e(n)=dn(n)-y(n);
    w=w+2*mu*sign(e(n))*x1/(ep+x1*x1');
    w1(n-M+1,:)=w(1,:);
end;
J=e.^2;
%the columns of w1 depict the history of the filter
%coefficients;
```

Signed-regressor algorithm

The signed-regressor or data-sign algorithm is given as follows

$$\mathbf{w}(n+1) = \mathbf{w}(n) + 2\mu\, e(n)\, sign(\mathbf{x}(n)) \tag{8.1.4}$$

where the sign function is applied to $\mathbf{x}(n)$ on element by element basis.

Sign-sign algorithm

The sign-sign algorithm is given by

$$\mathbf{w}(n+1) = \mathbf{w}(n) + 2\mu\, sign(e(n))\, sign(\mathbf{x}(n)) \tag{8.1.5}$$

Book MATLAB function for sign-sign algorithm

```
function[w,y,e,J,w1]=aalmssignsign(x,dn,mu,M)
%function[w,y,e,J,w1]=aalmssignsign(x,dn,mu,M)
%all quantities are real-valued;
%x=input data to the filter;dn=desired signal;
%M=order of the filter;
%mu=step-size parameter;x and dn must be of the same length
N=length(x);
y=zeros(1,N);
w=zeros(1,M);%initialized filter coefficient vector
for n=M:N
   x1=x(n:-1:n-M+1);%for each n the vector x1 is produced
    %of length M with elements from x in reverse order;
    y(n)=w*x1';
    e(n)=dn(n)-y(n);
    w=w+2*mu*sign(e(n))*sign(x1);
    w1(n-M+1,:)=w(1,:);
end;
J=e.^2;
%the columns of w1 depict the history of the filter
%coefficients;
```

8.2 Normalized LMS (NLMS) algorithm

Consider the LMS recursion algorithm

$$\mathbf{w}(n+1) = \mathbf{w}(n) + 2\mu(n)e(n)\mathbf{x}(n) \tag{8.2.1}$$

where the step-size parameter $\mu(n)$ varies in time. We have observed in Chapter 7 that the stability, convergence, and steady-state behavior of the LMS algorithm, are influenced by the filter length and the power of the signal.

Therefore, we can set

$$\mu(n) = \frac{1}{2\mathbf{x}^{\mathrm{T}}(n)\mathbf{x}(n)} = \frac{1}{2\|\mathbf{x}(n)\|^2} \tag{8.2.2}$$

in (8.2.1) to find the recursion

$$\mathbf{w}(n+1) = \mathbf{w}(n) + \frac{1}{\mathbf{x}^{\mathrm{T}}(n)\mathbf{x}(n)} e(n)\mathbf{x}(n) \tag{8.2.3}$$

The above equation can be found using a posteriori error (see Farhang-Boroujeny, 1999) minimization (see Problem 8.2.1) or using a constrained optimization procedure (see Haykin, 2001). Although (8.2.3) is appealing, in practice a more relaxed recursion is used that guarantees reliable results. Hence, we write

$$\mathbf{w}(n+1) = \mathbf{w}(n) + \frac{\bar{\mu}}{\varepsilon + \mathbf{x}^{\mathrm{T}}(n)\mathbf{x}(n)} e(n)\mathbf{x}(n) \tag{8.2.4}$$

where $\bar{\mu}$ and ε are constants. The small constant ε prevents division by a very small number of the data norm.

The normalized LMS (NLMS) algorithm is shown in Table 8.2.1.

Book MATLAB function for LMS algorithm with complex data

```
function[w,y,e,J,w1]=aacomplexnlms(x,dn,mubar,M,c)
%function[w,y,e,J,w1]=aacomplexnlms(x,dn,mubar,M,c)
%x=input data to the filter;dn=desired signal;
%M=filter order;c=constant;mubar=step-size equivalent
    %parameter;
%x and dn must be of the same length;J=learning curve;
N=length(x);
y=zeros(1,N);
w=zeros(1,M);%initialized filter coefficient vector;
for n=M:N
    x1=x(n:-1:n-M+1);%for each n vector x1 is of length
    %M with elements from x in reverse order;
    y(n)=conj(w)*x1';
    e(n)=dn(n)-y(n);
    w=w+(mubar/(c+conj(x1)*x1'))*conj(e(n))*x1;
    w1(n-M+1,:)=w(1,:);
end;
J=e.^2;
%the columns of the matrix w1 depict the history of the
%filter coefficients;
```

Table 8.2.1 The NLMS Algorithm

Real-valued functions		Complex-valued functions
Input:		
Initialization vector:	$\mathbf{w}(n) = \mathbf{0}$	
Input vector:	$\mathbf{x}(n)$	
Desired output:	$d(n)$	
Step-size parameter:	$\bar{\mu}$	
Constant:	ε	
Filter length:	M	
Output:		
Filter output:	$y(n)$	
Coefficient vector:	$\mathbf{w}(n+1)$	
Procedure:		
1) $y(n) = \mathbf{w}^{\mathrm{T}}(n)\mathbf{x} = \mathbf{w}(n)\mathbf{x}^{\mathrm{T}}(n)$		1) $y(n) = \mathbf{w}^{\mathrm{H}}(n)\mathbf{x}(n)$
2) $e(n) = d(n) - y(n)$		2) $e(n) = d(n) - \mathbf{w}^{\mathrm{H}}(n)\mathbf{x}(n)$
3) $\mathbf{w}(n+1) = \mathbf{w}(n) + \dfrac{\bar{\mu}}{\varepsilon + \mathbf{x}^{\mathrm{T}}(n)\mathbf{x}(n)} e(n)\mathbf{x}(n)$		3) $\mathbf{w}(n+1) = \mathbf{w}(n) + \dfrac{\bar{\mu}}{\varepsilon + \mathbf{x}^{\mathrm{H}}(n)\mathbf{x}(n)} \mathbf{x}(n)e^*(n)$

Note: The superscript H stands for Hermitian, or equivalently conjugate transpose.

8.3 Variable step-size LMS (VSLMS) algorithm

The VSLMS algorithm was introduced in 1986 to facilitate the conflicting requirements, whereas a large step-size parameter is needed for fast convergence and a small step-size parameter is needed to reduce the misadjustment factor. When the adaptation begins and $\mathbf{w}(n)$ is far from its optimum value, the step-size parameter should be large in order for the convergence to be rapid. However, as the filter coefficients $\mathbf{w}(n)$ approaches the steady-state solution, the step-size parameter should decrease in order to reduce the excess MSE.

To accomplish the variation of the step-size parameter, each filter coefficient is given a separate time-varying step-size parameter such that the LMS recursion algorithm takes the form

$$w_i(n+1) = w_i(n) + 2\mu_i(n)e(n)x(n-i) \quad i = 0, 1, \ldots, M-1 \tag{8.3.1}$$

where $w_i(n)$ is the ith coefficient of $\mathbf{w}(n)$ at iteration n and $\mu_i(n)$ is its associated step-size. The step-sizes are determined in an ad hoc manner, based on monitoring sign changes in the instantaneous gradient estimate $e(n)x(n-i)$. It was argued, that successive changes in the sign of the gradient estimate, indicates that the algorithm is close to its optimal solution and, hence, the step-size value must be decreased. The reverse is also true. The decision of decreasing the value of the step-size by some factor c_1 is based on some number m_1 successive changes of $e(n)x(n-i)$. Increasing the step-size parameter by some factor c_2 is based on m_2 successive sign changes. The parameters m_1 and m_2 can be adjusted to optimize performance, as can the factors c_1 and c_2.

Table 8.3.1 The VSLMS Algorithm

Input:

Initial coefficient vector: $\mathbf{w}(0)$
Input data vector: $\mathbf{x}(n) = [x(n)\ x(n-1)\ \ldots\ x(n-M+1)]^T$
Gradient term: $g_0(n-1) = e(n-1)x(n-1)$, $g_1(n-1) = e(n-1)x(n-1)$, ..., $g_{M-1}(n-1) = e(n-1)x(n-M)$

Step-size parameter: $\mu_0(n-1), \mu_1(n-1), \ldots, \mu_{M-1}(n-1)$
a = small positive constant
μ_{max} = positive constant

Outputs:

Desired output: $d(n)$
Filter output: $y(n)$
Filter update: $\mathbf{w}(n+1)$
Gradient term: $g_0(n), g_1(n), \ldots, g_{M-1}(n)$
Update step-size parameter: $\mu_0(n), \mu_1(n), \ldots, \mu_{M-1}(n)$

Execution:

1) $y(n) = \mathbf{w}^T(n)\mathbf{x}(n)$
2) $e(n) = d(n) - y(n)$
3) Weights and step-size parameter adaptation:
For $i = 0, 1, 2, \ldots, M-1$
$g_i(n) = e(n)x(n-i)$
$\mu_i(n) = \mu_i(n-1) + a\text{sign}(g_i(n))\textit{sign}(g_i(n))$
if $\mu_i(n) > \mu_{max}$, $\mu_i(n) = \mu_{max}$
if $\mu_i(n) < \mu_{min}$, $\mu_i = \mu_{min}$
$w_i(n+1) = w_i(n) + 2\mu_i(n)g_i(n)$
end

The set of update Equations (8.3.1) may be written in the matrix form

$$\mathbf{w}(n+1) = \mathbf{w}(n) + 2\mu(n)e(n)\mathbf{x}(n) \tag{8.3.2}$$

where $\mu(n)$ is a diagonal matrix with the following elements in its diagonal: $\mu_0(n), \mu_1(n), \cdots, \mu_{M-1}(n)$. The VSLMS algorithm is given in Table 8.3.1.

8.4 The leaky LMS algorithm

Let us assume that $\mathbf{x}(n)$ and $d(n)$ are jointly wide-sense stationary processes that determine when the coefficients $\mathbf{w}(n)$ converge in the mean to $\mathbf{w}^o = \mathbf{R}_x^{-1}\mathbf{p}_{dx}$. That is

$$\lim_{n\to\infty} E\{\mathbf{w}(n)\} = \mathbf{w}^o = \mathbf{R}_x^{-1}\mathbf{p}_{dx} \tag{8.4.1}$$

We start by taking the expectation of both sides of the LMS recursion as follows:

$$E\{\mathbf{w}(n+1)\} = E\{\mathbf{w}(n)\} + 2\mu E\{d(n)\mathbf{x}(n)\} - 2\mu E\{\mathbf{x}(n)\mathbf{x}^T(n)\mathbf{w}(n)\} \tag{8.4.2}$$

where $y(n) = \mathbf{x}^T(n)\mathbf{w}(n)$. Assuming that $\mathbf{x}(n)$ and $\mathbf{w}(n)$ are statistically independent (*independence* assumption), (8.4.2) becomes

$$\begin{aligned} E\{\mathbf{w}(n+1)\} &= E\{\mathbf{w}(n)\} + 2\mu E\{d(n)\mathbf{x}(n)\} - 2\mu E\{\mathbf{x}(n)\mathbf{x}^T(n)\}E\{\mathbf{w}(n)\} \\ &= (\mathbf{I} - 2\mu\mathbf{R}_{\mathbf{x}})E\{\mathbf{w}(n)\} + 2\mu\mathbf{p}_{\mathbf{dx}}(n) \end{aligned} \tag{8.4.3}$$

which is similar to the steepest-descent method equation, see (6.2.3), with the difference that here we have ensemble average. This suggests that the steepest-descent method is applicable to ensemble average $E\{\mathbf{w}(n+1)\}$. Rewriting (6.2.8) in its matrix form we obtain

$$\xi'(n) = (\mathbf{I} - 2\mu\Lambda)^n \xi'(0) \quad k = 0,1,\cdots,M-1; \quad n = 1,2,3,\cdots \tag{8.4.4}$$

We observe that $\mathbf{w}(n) \to \mathbf{w}^o$ if $\xi(n) = \mathbf{w}(n) - \mathbf{w}^o \to 0$ as $n \to \infty$ or when $\xi'(n) = \mathbf{Q}^T\xi(n)$ converges to zero. The kth row of (8.4.4) is

$$\xi'_k(n) = (1 - 2\mu\lambda_k)^n \xi'_k(0) \tag{8.4.5}$$

which indicates that $\xi'_k(n) \to 0$ if

$$|1 - 2\mu\lambda_k| < 1 \quad k = 0,1,\cdots,M-1 \tag{8.4.6}$$

or

$$0 < \mu < \frac{1}{\lambda_k} \tag{8.4.7}$$

To be more restrictive, we must set the value of μ to obey the inequality

$$0 < \mu < \frac{1}{\lambda_{max}} \tag{8.4.8}$$

If $\lambda_k = 0$, (8.4.5) indicates that no convergence takes place as n approaches infinity. Since it is possible for these undamped modes to become unstable, it is important for the stabilization of the LMS algorithm to force these modes to zero. One way to avoid this difficulty is to introduce a leakage coefficient into the LMS algorithm as follows:

$$\mathbf{w}(n+1) = (1 - 2\mu\gamma)\mathbf{w}(n) + 2\mu e(n)\mathbf{x}(n) \tag{8.4.9}$$

where $0 < \gamma << 1$. The effect of introducing the *leakage coefficient* γ is to force any undamped modes to become zero and to force the filter coefficients to zero if either $e(n)$ or $\mathbf{x}(n)$ is zero (the homogeneous equation $w_i(n+1) = (1 - 2\mu\gamma)w_i(n)$ has the solution $w_i(n) = A(1 - 2\mu\gamma)^n$, where A is a constant).

We write (8.4.9) in the form $(e(n) = d(n) - \mathbf{x}^T(n)\mathbf{w}(n))$

$$\mathbf{w}(n+1) = [\mathbf{I} - 2\mu[\mathbf{x}(n)\mathbf{x}^T(n) + \gamma\mathbf{I}]\mathbf{w}(n) + 2\mu d(n)\mathbf{x}(n) \tag{8.4.10}$$

By taking the expected value of both sides of the above equation and using the independence assumption, we obtain

$$E\{\mathbf{w}(n+1)\} = [\mathbf{I} - 2\mu[\mathbf{R}_x + \gamma\mathbf{I}]E\{\mathbf{w}(n)\} + 2\mu\mathbf{p}_{dx}(n) \quad (8.4.11)$$

Comparing the above equation with (8.4.3), we observe that the autocorrelation matrix $\mathbf{R}_x$ of the LMS algorithm has been replaced with $\mathbf{R}_x + \gamma\mathbf{I}$. Since the eigenvalues of $\mathbf{R}_x + \gamma\mathbf{I}$ are $\lambda_k + \gamma$, and since $\lambda_k \geq 0$, all the modes of the leaky LMS algorithm will be decayed to zero. Furthermore, the constraint for the step-size parameter becomes

$$0 < \mu < \frac{1}{\lambda_{\max} + \gamma} \quad (8.4.12)$$

As $n \to \infty$, $\mathbf{w}(n+1) \cong \mathbf{w}(n)$ and, hence, (8.4.11) becomes

$$\lim_{n\to\infty} E\{\mathbf{w}(n)\} = [\mathbf{R}_x + \gamma\mathbf{I}]^{-1}\mathbf{p}_{dx} \quad (8.4.13)$$

which indicates that the leakage coefficient introduces a bias into steady-state solution $\mathbf{R}_x^{-1}\mathbf{p}_{dx}$. For another way to produce the leaky LMS algorithm see Problem 7.4.3.

Book leaky LMS MATLAB function

```
function[w,y,e,J,w1]=aaleakylms(x,dn,mu,gama,M)
%function[w,y,e,J,w1]=aaleakylms(x,dn,mu,gama,M)
%all signals are real valued;x=input to filter;
%y=output from the filter;dn=desired signal;
%mu=step-size factor;gama=gamma factor<<1;
%M=number of filter coefficients;w1=matrix whose M
%rows give the history of each filter coefficient;
N=length(x);
y=zeros(1,N);
w=zeros(1,M);
for n=M:N
      x1=x(n:-1:n-M+1);
      y(n)=w*x1';
      e(n)=dn(n)-y(n);
      w=(1-2*mu*gama)*w+2*mu*e(n)*x1;
      w1(n-M+1,:)=w(1,:);
end;
J=e.^2;
```

Figure 8.4.1 shows the input data $\{x(n)\}$ to the leaky LMS filter, the output of the filter and the learning curve. The following signals and parameters were used: dn = desired signal = $sin(0.1\pi n)$, v = 2(rand-0.5), x = data = dn + v, $\mu = 0.01$, $\gamma = 0.1$, $M = 16$.

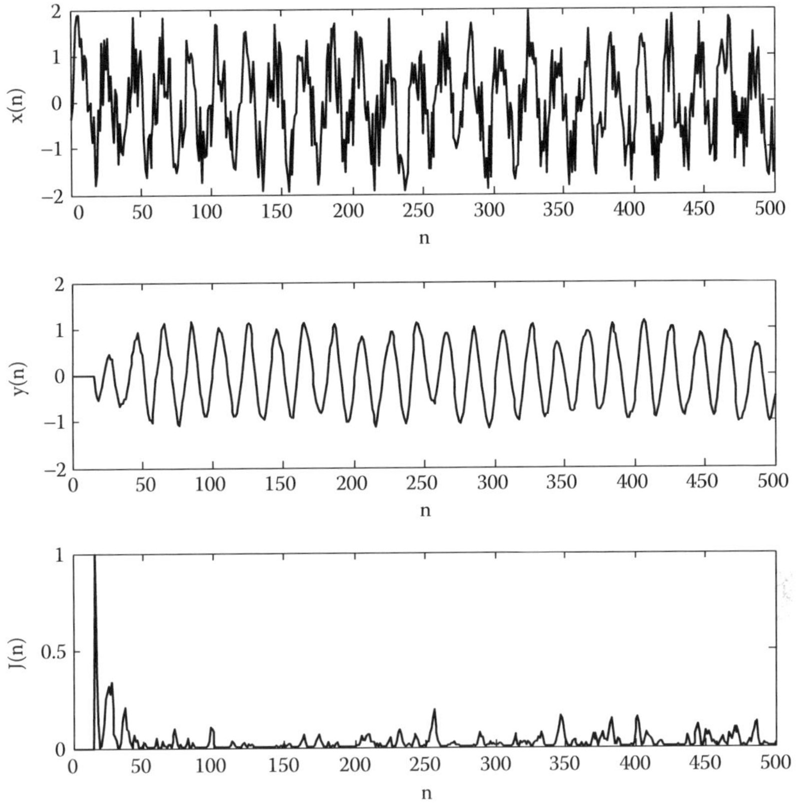

Figure 8.4.1 The input data, the output signal, and the learning curve of an adaptive filtering problem using leaky LMS algorithm.

8.5 Linearly constrained LMS algorithm

In all previous analyses of Wiener filtering problem, steepest-descent method, Newton's method, and the LMS algorithm, no constrain was imposed on the solution of minimizing the MSE. However, in some applications there might be some mandatory constraints that must be taken into consideration in solving optimization problems. For example, the problem of minimizing the average output power of a filter while the frequency response must remain constant at specific frequencies (Haykin, 2001). In this section, we discuss the filtering problem of minimizing the MSE subject to a general constraint.

The error between the desired signal and the output of the filter is

$$e(n) = d(n) - \mathbf{w}^{\mathrm{T}}(n)\mathbf{x}(n) \tag{8.5.1}$$

We wish to minimize this error in the mean-square sense, subject to the constraint

$$\mathbf{c}^{\mathrm{T}}\mathbf{w} = a \tag{8.5.2}$$

where a is a constant and $\mathbf{c}$ is a fixed vector. Using the Lagrange multiplier method, we write

$$J_c = E\{e^2(n)\} + \lambda(\mathbf{c}^{\mathrm{T}}\mathbf{w} - a) \tag{8.5.3}$$

where λ is the Lagrange multiplier. Hence, the following relations

$$\nabla_{\mathbf{w}} J_c = 0 \quad \text{and} \quad \frac{\partial J_c}{\partial \lambda} = 0 \tag{8.5.4}$$

must be satisfied simultaneously. The term $\partial J_c/\partial\lambda$ produces the constraint (8.5.2). Next we substitute the error $e(n)$ in (8.5.3) to obtain (see Problem 8.5.1)

$$J_c = J_{\min} + \xi^{\mathrm{T}}\mathbf{R}_{\mathbf{x}}\xi + \lambda(\mathbf{c}^{\mathrm{T}}\xi - a') \tag{8.5.5}$$

where

$$\xi(n) = \mathbf{w}(n) - \mathbf{w}^{\mathrm{o}}, \quad \mathbf{w}^{\mathrm{o}} = \mathbf{R}_{\mathbf{x}}^{-1}\mathbf{p}_{\mathbf{dx}}, \quad \mathbf{R}_{\mathbf{x}} = E\{\mathbf{x}(n)\mathbf{x}^{\mathrm{T}}(n)\} \tag{8.5.6}$$

and

$$\mathbf{p}_{\mathbf{dx}} = E\{d(n)\mathbf{x}(n)\}, \quad a' = a - \mathbf{c}^{\mathrm{T}}\mathbf{w}^{\mathrm{o}} \tag{8.5.7}$$

The solution now has changed to $\nabla_{\xi} J = 0$ and $\partial J/\partial\lambda = 0$. Hence, from (8.5.5) we obtain

$$\nabla_{\xi} J_c = \begin{bmatrix} \dfrac{\partial J_c}{\partial \xi_1} \\ \vdots \\ \dfrac{\partial J_c}{\partial \xi_M} \end{bmatrix} = \begin{bmatrix} 2\xi_1 r_1 + 2\xi_2 r_2 + \cdots + 2\xi_M r_M \\ \vdots \qquad\qquad \vdots \\ 2\xi_1 r_M + 2\xi_2 r_{M-1} + \cdots 2\xi_M r_1 \end{bmatrix} + \lambda \begin{bmatrix} c_1 \\ \vdots \\ c_M \end{bmatrix} = \mathbf{0} \tag{8.5.8}$$

or in matrix form

$$2\mathbf{R}_{\mathbf{x}}\xi_c^o + \lambda\mathbf{c} = \mathbf{0} \tag{8.5.9}$$

where ξ_c^o is the constraint optimum value of the vector ξ. In addition, the constraint gives the relation

$$\frac{\partial J_c}{\partial \lambda} = \mathbf{c}^T \xi_c^o - a' = 0 \tag{8.5.10}$$

Solving the system of the last two equations for λ and ξ_c^o we obtain

$$\lambda = -\frac{2a'}{\mathbf{c}^T \mathbf{R}_x^{-1} \mathbf{c}} \qquad \xi_c^o = \frac{a' \mathbf{R}_x^{-1} \mathbf{c}}{\mathbf{c}^T \mathbf{R}_x^{-1} \mathbf{c}} \tag{8.5.11}$$

Substituting the value of λ in (8.5.5) we obtain the minimum value of J_c to be

$$J_c = J_{\min} + \frac{a'^2}{\mathbf{c}^T \mathbf{R}_x^{-1} \mathbf{c}} \tag{8.5.12}$$

But $\mathbf{w}(n) = \xi(n) + \mathbf{w}^o$ and, hence, using (8.5.11) we obtain the equation

$$\mathbf{w}_c^o = \mathbf{w}^o + \frac{a' \mathbf{R}_x^{-1} \mathbf{c}}{\mathbf{c}^T \mathbf{R}_x^{-1} \mathbf{c}} \tag{8.5.13}$$

Note: The second term of (8.5.12) is the excess MSE produced by the constraint.

To obtain the recursion relation subject to constraint (8.5.2), we must proceed in two steps:

Step 1:

$$\mathbf{w}'(n) = \mathbf{w}(n) + 2\mu e(n)\mathbf{x}(n) \tag{8.5.14}$$

Step 2:

$$\mathbf{w}(n+1) = \mathbf{w}'(n) + \eta(n) \tag{8.5.15}$$

where $\eta(n)$ is chosen so that $\mathbf{c}^T \mathbf{w}(n+1) = a$ while $\eta^T(n)\eta(n)$ is minimized. In other words, we choose the vector $\eta(n)$ so that (8.5.2) holds after Step 2, while the perturbation introduced by $\eta(n)$ is minimized. The problem can be solved using the Lagrange multiplier method that gives

$$\eta(n) = \frac{a - \mathbf{c}^T \mathbf{w}'(n)}{\mathbf{c}^T \mathbf{c}} \mathbf{c} \tag{8.5.16}$$

Thus, we obtain the final form of (8.5.15) to be

$$\mathbf{w}(n+1) = \mathbf{w}'(n) + \frac{a - \mathbf{c}^{\mathrm{T}}\mathbf{w}'(n)}{\mathbf{c}^{\mathrm{T}}\mathbf{c}} \tag{8.5.17}$$

The constraint algorithm is given in Table 8.5.1.

Table 8.5.1 Linearly Constrained LMS Algorithm

Input:	Initial coefficient vector, $\mathbf{w}(0) = \mathbf{0}$
	Input data vector, $\mathbf{x}(n)$
	Desired output, $d(n)$
	Constant vector, $\mathbf{c}$
	Constraint constant, a
Output:	Filter output, $y(n)$
Procedure:	$y(n) = \mathbf{w}^{\mathrm{T}}(n)\mathbf{x}(n)$
	$e(n) = d(n) - y(n)$
	$\mathbf{w}'(n) = \mathbf{w}(n) + 2\mu e(n)\mathbf{x}(n)$
	$\mathbf{w}(n+1) = \mathbf{w}'(n) + \frac{a - \mathbf{c}^{\mathrm{T}}\mathbf{w}'(n)}{\mathbf{c}^{\mathrm{T}}\mathbf{c}}\mathbf{c}$

Book constraint LMS MATLAB function

```
function[w,e,y,J,w2]=aaconstrainedlms(x,dn,c,a,mu,M)
%function[w,e,y,J,w2]=aaconstrainedlms(x,dn,c,a,mu,M);
%x=data vector;dn=desired vector of equal length with x;
%c=constant row vector of length M;a=constant, e.g.
%a=0.8;mu=step-
%size parameter;M=filter order(number of filter
%coefficients);
%w2=matrix whose columns give the history of each
%coefficient;
w=zeros(1,M);
N=length(x);
for n=M:N;
    y(n)=w*x(n:-1:n-M+1)';
    e(n)=dn(n)-y(n);
    w1=w+2*mu*e(n)*x(n:-1:n-M+1);
    w=w1+((a-c*w1')*c/(c*c'));
    w2(n-M+1,:)=w(1,:);
end;
J=e.^2;
```

Figure 8.5.1 shows the results of a constrained LMS filter with the following data: dn = $sin(0.1n\pi)$; v = noise = 2(rand-0.5); x = data = dn + v; c = ones(1, 32); $a = 0.8$; $\mu = 0.01$; $M = 32$.

As an example of solving a constrained optimization problem using Lagrange multiplier method, the NLMS recursion can be obtained as a solution of the following problem:

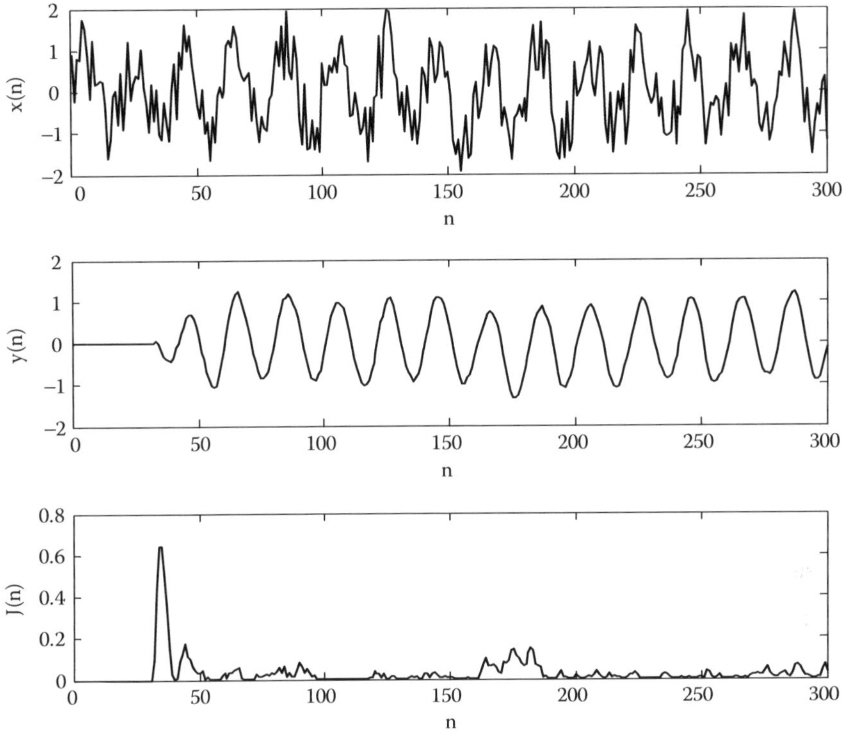

Figure 8.5.1 The input data, the output signal, and the learning curve of an adaptive filtering problem using constrained LMS algorithm.

Minimize $\min_{\mathbf{w}} \|\mathbf{w}(n)-\mathbf{w}(n-1)\|^2$ subject to the constraint $d(n)=\mathbf{w}^{\mathrm{T}}(n+1)\mathbf{x}(n)$

The first step in the solution is to write the cost function as:

$$J(n)=\left\|\Delta\mathbf{w}\right\|^2+\lambda[d(n)-\mathbf{w}(n+1)x(n)] \tag{8.5.18}$$

where

$$\Lambda\mathbf{w}=\mathbf{w}(n+1)-\mathbf{w}(n) \tag{8.5.19}$$

Differentiating the cost function with respect to $\mathbf{w}(n+1)$ leads to

$$\frac{\partial J(n)}{\partial\mathbf{w}(n+1)}=2(\mathbf{w}(n+1)-\mathbf{w}(n))-\lambda\,\mathbf{x}(n) \tag{8.5.20}$$

Setting this results to zero results in

$$\mathbf{w}(n+1)=\mathbf{w}(n)+\frac{1}{2}\lambda\,\mathbf{x}(n) \tag{8.5.21}$$

Substituting this last result into the constraint $d(n) = \mathbf{w}^T(n+1)\mathbf{x}(n)$, we obtain

$$d(n) = \left(\mathbf{w}(n) + \frac{1}{2}\lambda\mathbf{x}(n)\right)^T \mathbf{x}(n)$$
$$= \mathbf{w}^T(n)\mathbf{x}(n) + \frac{1}{2}\lambda\|\mathbf{x}(n)\|^2 \tag{8.5.22}$$

Since $e(n) = d(n) - \mathbf{w}^T(n)\mathbf{x}(n)$, solving Equation (8.5.22) for λ leads to

$$\lambda = \frac{2e(n)}{\|\mathbf{x}(n)\|^2} \tag{8.5.23}$$

Substituting Equation (8.5.23) in (8.5.21) results in

$$\mathbf{w}(n+1) = \mathbf{w}(n) + \frac{1}{\|\mathbf{x}(n)\|^2} e(n)\mathbf{x}(n) \tag{8.5.24}$$

Finally, introducing a factor μ in Equation (8.5.24) to control the change in the weight vector, we obtain

$$\mathbf{w}(n+1) = \mathbf{w}(n) + \frac{\mu}{\|\mathbf{x}(n)\|^2} e(n)\mathbf{x}(n) \tag{8.5.25}$$

Clearly, Equation (8.5.25) is the conventional NLMS algorithm.

8.6 Self-correcting adaptive filtering (SCAF)

One way by which we may improve the output of the adaptive filter so that it is approximately equal to the desired one is to use a *self-correcting adaptive filtering* as shown in Figure 8.6.1. In this proposed configuration the desired

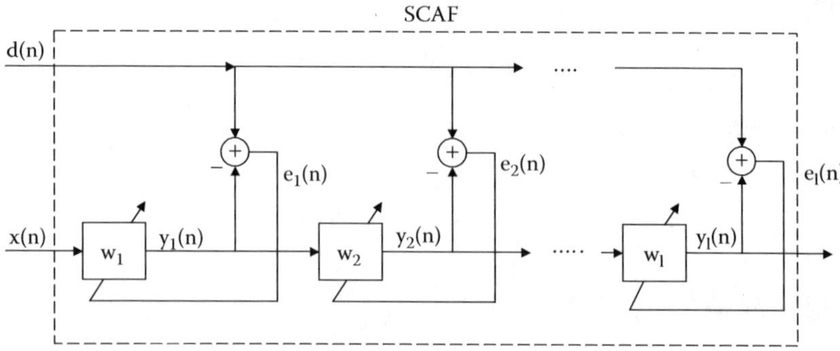

Figure 8.6.1 Block diagram of the SCAF method.

signal is compared with signals that become closer and closer to the desired one. The output of the ith stage is related to the previous one as follows:

$$y_{i+1}(n) = y_i * w_{i+1} \tag{8.6.1}$$

Book MATLAB function for self-correcting LMS algorithm

```
function[w,y,e,J]=aaselfcorrectinglms(x,dn,mu,M,I)
%function[w,y,e,J]aaselfcorrectinglms(x,dn,mu,M,I);
[w(1,:),y(1,:),e(1,:)]=aalms(x,dn,mu,M);
for i=2:I%I=number of iterations, I<8-10 is sufficient;
    [w(i,:),y(i,:),e(i,:)]=aalms(y(i-1,:),dn,mu,M);
end;
J=e.^2;
```

Figure 8.6.2a shows the input data into the self-correcting filter, Figure 8.6.2b shows the output, $y_1(n)$, of the first stage of the self-correcting filter, and Figure 8.6.2c shows the output, $y_{10}(n)$, of the filter at its tenth stage. The data for these results were: dn = desired signal = $sin(0.1n\pi)$; v = noise = randn; x = input data = dn + v; mu = 0.01; M = 10, I = number of stages = 10.

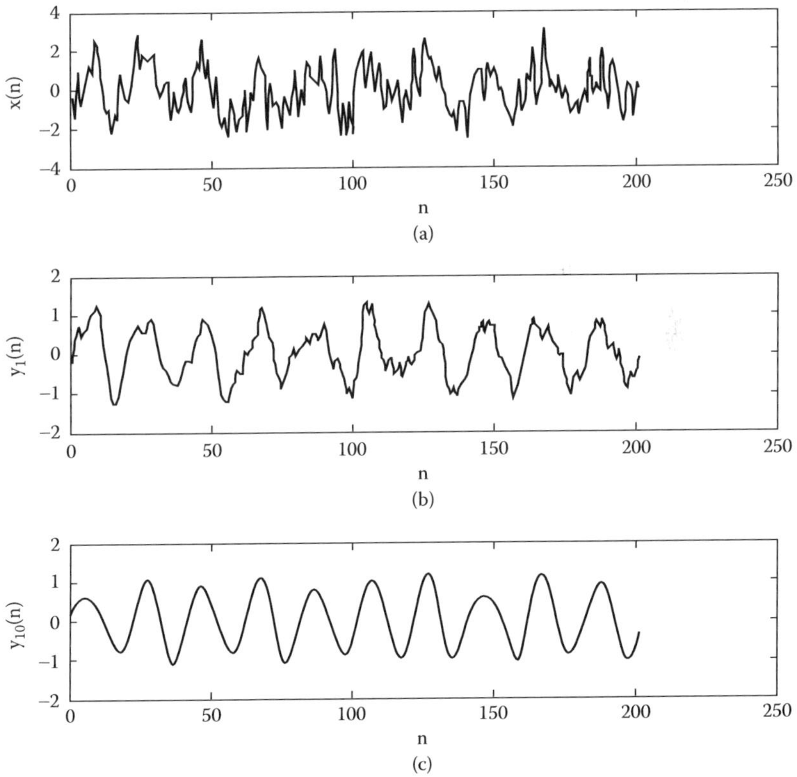

Figure 8.6.2 Noise reduction with a self-correcting filter.

Book MATLAB function for the self-correcting sign regressor

```
function[w,y,e,J]=aaselfcorsignregreslms(x,dn,mu,M,I)
%function[w,y,e,J]=aaselfcorsignregreslms(x,dn,mu,M,I);
%x=input data to the filter;dn=desired sig-
nal;length(x)=length(dn);
%y=output of the filter an Ixlength(x) matrix;J=error
%function an
%Ixlength(x) matrix;I=number of stages;
[w(1,:),y(1,:),e(1,:),J(1,:)]=aalmssignedregressor...
(x,dn,mu,M);
for i=2:I
    [w(i,:),y(i,:),e(i,:),J(i,:)]=aalms(y(i-1,:),dn,mu,M);
end;
J=e.^2;
```

Figure 8.6.3 shows the learning curves for the output of the first and fourth stages. It is apparent that the self-correcting adaptive filter improves with the number of stages used. The data were: dn = desired signal = $sin(0.2n\pi)$, v = noise = 1.5(rand-0.5), x = dn + v, $\mu = 0.01$, $M = 8$, and I = 4.

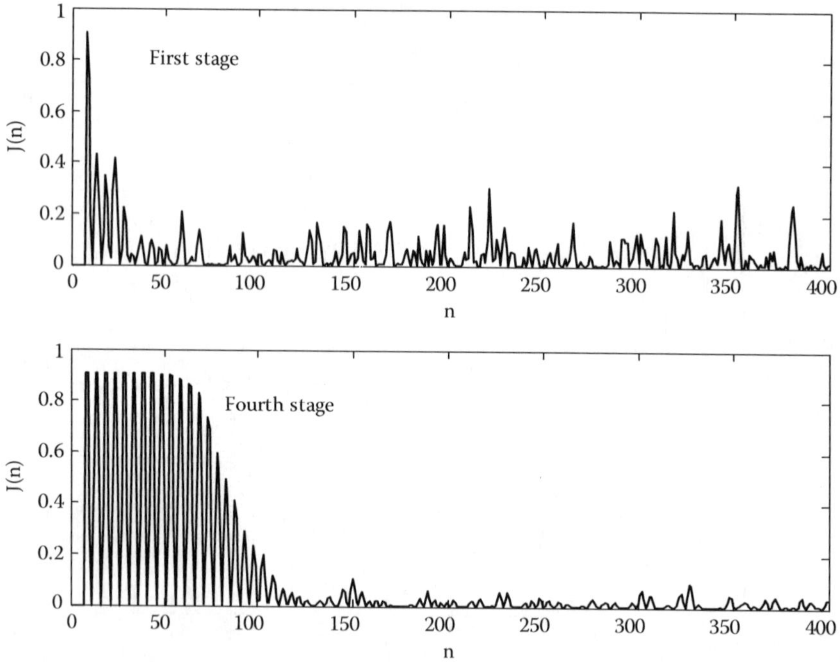

Figure 8.6.3 Learning curves of the first and fourth stages of a self-correcting filter.

Book MATLAB function for self-correcting sign-sign algorithm

```
function[w,y,e,J]=aaselfcorsignsignlms(x,dn,mu,M,I)
%function[w,y,e,J]=aaselfcorsignsignlms(x,dn,mu,M,I);
%x=input data to the filter;y=output data from the filter,
%y is an Ixlength(x) matrix; J=learning curves, an Ix-
%length(x)
%matrix;mu=step-size parameter;M=umber of coefficients;I=
%number of stages;w=an Ixlength(x) matrix of filter coef-
%ficients;
%dn=desired signal;
[w(1,:),y(1,:),e(1,:),J(1,:)]=aalmssignsign(x,dn,mu,M);
for i=2:I
   [w(i,:),y(i,:),e(i,:),J(i,:)]=aalms(y(i-1,:),dn,mu,M);
end;
J=e.^2;
```

The self-correcting adaptive filtering can easily be implemented by using all types of adaptive filters such as normalized, constrained, transform domain, etc.

8.7 Transform domain adaptive LMS filtering

The implementation of the LMS filter in the frequency domain can be accomplished simply by taking the Discrete Fourier Transform (DFT) of both the input data, $\{x(n)\}$, and the desired signal, $\{d(n)\}$. The advantage of doing this is due to the fast processing of the signal using the Fast Fourier Transform (fft) algorithm. However, this procedure requires a block-processing strategy, which results in storing a number of incoming data in buffers, and thus, some delay is unavoidable.

The simplest approach is that given by Dentino et al. (1978), and it is shown in Figure 8.7.1. The signals are processed by block-by-block format, that is $\{x(n)\}$ and $\{d(n)\}$ are sequenced into blocks of length M so that

$$x_i(n) = x(iM+n), \quad d_i(n) = d(iM+n) \quad n = 0, 1, \cdots, M-1; \quad i = 0, 1, \cdots \tag{8.7.1}$$

The values of the ith block of the signals $\{x_i(n)\}$ and $\{d_i(n)\}$ are Fourier transformed using the DFT to find $X_i(k)$ and $D_i(k)$, respectively. Due to DFT properties, the sequences $X_i(k)$ and $D_i(k)$ have M complex elements corresponding to frequency indices ('bins') $k = 0, 1, \ldots, M-1$

$$X_i(k) = DFT\{x_i(n)\} = \sum_{n=0}^{M-1} x_i(n) e^{-j\frac{2\pi nk}{M}} \quad k = 0, 1, \cdots, M-1 \tag{8.7.2}$$

$$D_i(k) = DFT\{d_i(n)\} = \sum_{n=0}^{M-1} d_i(n) e^{-j\frac{2\pi nk}{M}} \quad k = 0, 1, \cdots, M-1 \tag{8.7.3}$$

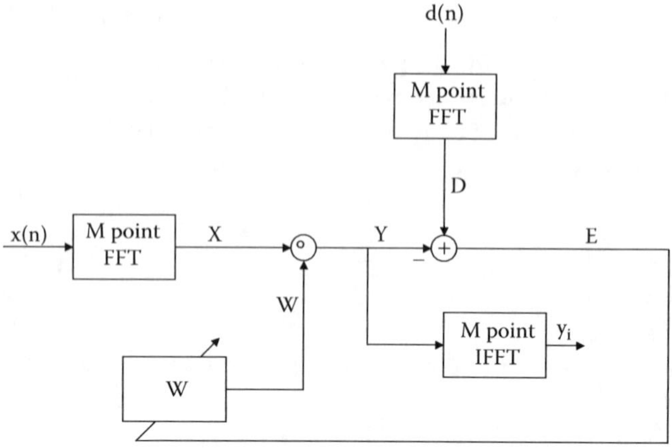

Figure 8.7.1 Frequency domain adaptive filter (the circle indicates multiplication term by term and the subscript *i* indicates the ith block of data).

During the ith block processing, the output in each frequency bin of the adaptive filter is computed by

$$Y_i(k) = W_{i,k} X_i(k) \quad k = 0,1,2,\cdots,M-1 \tag{8.7.4}$$

where $W_{i,k}$ is the kth frequency bin corresponding to the ith update (corresponding to the ith block data). The error in the frequency domain is

$$E_i(k) = D_i(k) - Y_i(k) \quad k = 0,1,2,\cdots,M-1 \tag{8.7.5}$$

The system output is given by

$$y_i(n) = y(iM+n) = IDFT\{Y_i(k)\} = \frac{1}{M}\sum_{k=0}^{M-1} Y_i(k) e^{j\frac{2\pi nk}{M}} \quad n = 0,1,2,\cdots M-1 \tag{8.7.6}$$

To update the filter coefficients we use, by analogy to LMS recursion, the following recursion:

$$\mathbf{w}_{i+1} = \mathbf{w}_i + 2\mu \mathbf{E}_i \bullet \mathbf{X}_i^* \tag{8.7.7}$$

where

$$\mathbf{W}_{i+1} = [W_{i+1,0} \quad W_{i+1,1} \quad \cdots \quad W_{i+1,M-1}]^T$$

$$\mathbf{W}_i = [W_{i,0} \quad W_{i,1} \quad \cdots \quad W_{i,M-1}]^T$$

$$\mathbf{E}_i = [E_i(0) \quad E_i(1) \quad \cdots \quad E_i(M-1)]^T$$

$$\mathbf{X}_i^* = [X_i^*(0) \quad X_i^*(1) \quad \cdots \quad X_i^*(M-1)]^T$$

The dot (.) in (8.7.7) implies element-by-element multiplication and * stands for complex conjugate. If we set X_i^* in the form

$$\mathbf{X}_\mathrm{i} = diag\{X_i(0) \quad X_i(1) \ \cdots \ X_i(M-1)\} = \begin{bmatrix} X_i(0) & 0 & \cdots & 0 \\ 0 & X_i(1) & \cdots & 0 \\ \vdots & \vdots & \cdots & 0 \\ 0 & 0 & & X_i(M-1) \end{bmatrix} \tag{8.7.8}$$

then (8.7.7) becomes

$$\mathbf{W}_{\mathrm{i+1}} = \mathbf{W}_\mathrm{i} + 2\mu \mathbf{X}_\mathrm{i}^* \mathbf{E}_\mathrm{i} \tag{8.7.9}$$

Therefore, the Equations (8.7.1)–(8.7.7) constitute the frequency domain of the LMS algorithm. The Book MATLAB function that gives the coefficients after I blocks (or iterations) is given below.

Book MATLAB Fourier transform LMS algorithm

```
function[A]=aaftlms(x,d,M,I,mu)
%function[A]=aaftlms(x,d,M,I,mu);
wk=zeros(1,M);
for i=0:I          %I=number of iterations (or blocks);
    if I*M>length(x)-1
        ('error:I*M<length(x)-1')
    end;
                        %M=number of filter coefficients;
        x1=x(M*(i+1):-1:i*M+1);
        d1=d(M*(i+1):-1:i*M+1);
    xk=fft(x1);
    dk=fft(d1);
    yk=wk.*xk;
    ek=dk-yk;
    wk=wk+2*mu*ek.*conj(xk);
    A(i+1,:)=wk;
end;
%all the rows of A are the wk's at an increase order
%of iterations(blocks);
%to filter the data, wk must be inverted in thc time
%domain, convolve with the data x and then plot the
%real part of the output y, e.g. wn4=the forth iteration
%=ifft(A(4,:)),yn4=filter(wn4/4,1,x) for even M;
```

Example 8.7.1: The following Book MATLAB program produced the outputs that are presented in Figure 8.7.2.

```
M=32; I=10; mu=0.01;
n=0:999;
d=sin(0.1*pi*n); v=1.5*randn(1,1000);
```

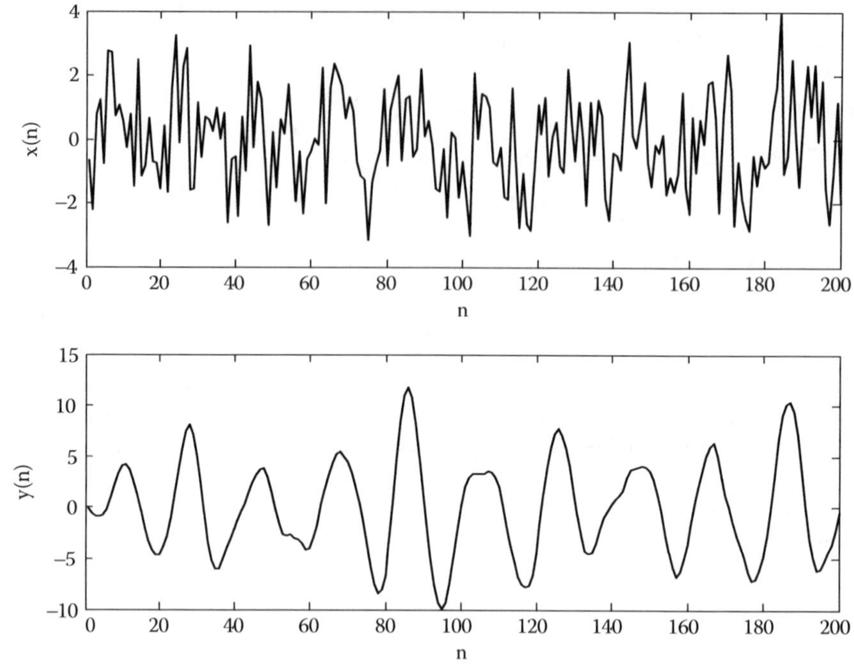

Figure 8.7.2 Adaptive filtering with FT LMS algorithm.

```
x=d+v;
[A]=aaftlms(x,d,M,I,mu);
wn8=ifft(A(8,:));% the inverse fft of row 8 of A;
yn8=filter(wn8,1,x);
subplot(2,1,1);plot(x(1,1:200));xlabel('n');ylabel...
    ('x(n)');
subplot(2,1,2);plot(real(yn8(1,1:200)));xlabel('n');...
    ylabel ('y(n)');
```

Convergence

Let the signals $x(n)$ and $y(n)$ be jointly stationary, and let the initial filter coefficient be zero, $\mathbf{W}_0 = \mathbf{0}$. Substituting (8.6.4) and (8.6.5) in, we obtain (see Problem 8.7.1)

$$W_{i+1,k} = (1 - 2\mu|X_i|^2)W_{i,k} + 2\mu D_i(k)X_i^*(k) \tag{8.7.10}$$

The expected value of (8.7.10), assuming $W_{i,k}$ and $X_i(k)$ are statistically independent, is given by

$$E\{W_{i+1,k}\} = (1 - 2\mu E\{|X_i(k)|^2\})E\{W_{i,k}\} + 2\mu E\{D_i(k)X_i^*(k)\} \tag{8.7.11}$$

Because $x(n)$ and $x(k)$ are stationary, their statistical characteristics do not change from block to block and, therefore, the ensembles $E\{|X_i(k)|^2\}$ and $E\{D_i(k)X_i^*(k)\}$ are independent of i but depend on k. Taking the Z-transform of (8.7.11) with respect to i of the dependent variable $W_{i,k}$, we find the relation (see Table 2.3.1 and Problem 8.7.2)

$$W_k(z) = -2\mu E\left\{\left|X_i(k)\right|^2\right\}\frac{W_k(z)}{z-1} + \frac{2\mu E\{D_i(k)X_i^*(k)\}}{z-1} \tag{8.7.12}$$

Applying the final value theorem (see Table 2.3.1 and Problem 8.7.2), we obtain the steady-state value for the filter coefficients

$$E\{W_k^\infty\} = \frac{E\{D_i(k)X_i^*(k)\}}{E\{\left|X_i(k)\right|^2\}} \tag{8.7.13}$$

Let the mean filter coefficient error $\mathrm{E}_i(k)$ be defined by

$$\mathrm{E}_\mathrm{i}(k) = E\{W_{i,k}\} - E\{W_k^\infty\} \tag{8.7.14}$$

Then, using (8.7.13) and (8.7.11), we find (see Problem 8.7.3)

$$\mathrm{E}_{i+1}(k) = (1 - 2\mu E\{\left|X_i(k)\right|^2\})\mathrm{E}_\mathrm{i}(k) \quad k = 0,1,2,\cdots,M-1 \tag{8.7.15}$$

Using the iteration approach for the solution of the above difference equation, we find

$$\mathrm{E}_\mathrm{i}(k) = (1 - E\{\left|X_i(k)\right|^2\})^i\,\mathrm{E}_0(k) \quad k = 0,1,2,\cdots,M-1 \tag{8.7.16}$$

The solution converges if

$$\left|1 - 2\mu E\{\left|X_i(k)\right|^2\}\right| < 1 \quad or \quad 0 < \mu < \frac{1}{E\{\left|X_i(k)\right|^2\}} \tag{8.7.17}$$

which shows that the power of the input plays a fundamental role in convergence and stability.

8.8 Error normalized LMS algorithms

A new class of LMS algorithms based on error normalization has been reported by the authors in IEEE conferences in 2004 and 2005. These algorithms are:

1. Error Normalized Step-Size (ENSS) LMS Algorithm

$$\mathbf{w}(n+1) = \mathbf{w}(n) + \frac{\mu}{1+\mu \left\| \mathbf{e}_{\mathrm{L}}(n) \right\|^2} \mathbf{x}(n)\, e(n) \tag{8.8.1}$$

2. Robust Variable Step-Size (RVSS) LMS Algorithm

$$\mathbf{w}(n+1) = \mathbf{w}(n) + \frac{\mu \left\| \mathbf{e}_{\mathrm{L}}(n) \right\|^2}{\alpha \left\| \mathbf{e}(n) \right\|^2 + (1-\alpha) \left\| \mathbf{x}(n) \right\|^2} \mathbf{x}(n) e(n) \tag{8.8.2}$$

3. Error-Data Normalized Step-Size (EDNSS) LMS Algorithm

$$\mathbf{w}(n+1) = \mathbf{w}(n) + \frac{\mu}{\alpha \left\| \mathbf{e}_{\mathrm{L}}(n) \right\|^2 + (1-\alpha) \left\| \mathbf{x}(n) \right\|^2} \mathbf{x}(n) e(n) \tag{8.8.3}$$

where

$$\left\| \mathbf{e}_{\mathrm{L}}(n) \right\|^2 = \sum_{i=0}^{\mathrm{L}-1} \left| e(n-i) \right|^2 \tag{8.8.4}$$

and

$$\left\| \mathbf{e}(n) \right\|^2 = \sum_{i=0}^{n-1} \left| e(n-i) \right|^2 \tag{8.8.5}$$

Comments

- The parameters α, L, and μ in all of these algorithms are appropriately chosen to achieve the best trade-off between rate of convergence and low final MSE. L could be constant or variable (L = n, for example), depending on whether the underlying environment is stationary or nonstationary.
- The variable step-sizes in all of these algorithms should vary between two predetermined hard limits. The lower value guarantees the capability of the algorithm to respond to an abrupt change that could happen at a very large value of iteration number n, while the maximum value maintains stability of the algorithm.

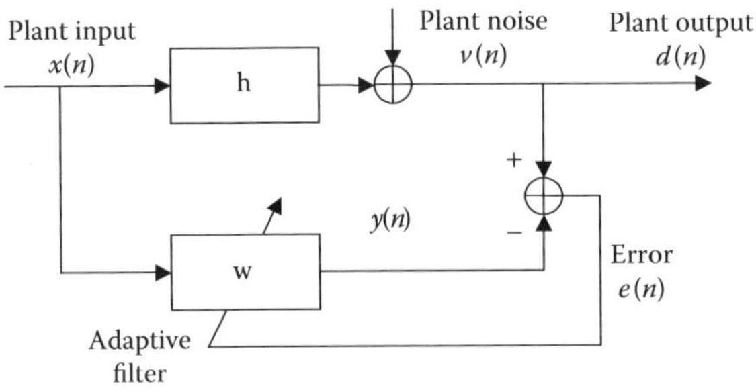

Figure 8.8.1 Adaptive plant identification.

Simulations

The ENSS algorithm using an error vector of increasing length (L = n) is compared first with the NLMS algorithm in system identification shown in Figure 8.8.1. The adaptive filter and the unknown system are both excited by a zero-mean white Gaussian signal of unit variance. The length of the unknown system impulse response is assumed to be N = 4. The internal unknown system noise $v(n)$ is assumed to be white Gaussian with mean = 0 and variance $\sigma_v^2 = 0.09$. The optimum values of μ in both algorithms are chosen to obtain the same exact value of misadjustment, $\mathcal{M} = 2\%$. The value of $\mathcal{M}$ is estimated by averaging excess mean-square error (EMSE) over n after the algorithm has reached steady-state, and dividing the result by σ_v^2. Simulation plots are obtained by ensemble averaging of 200 independent simulation runs. Figure 8.8.2 shows the learning curves of the two algorithms. While retaining the same level of misadjustment, the ENSS algorithm

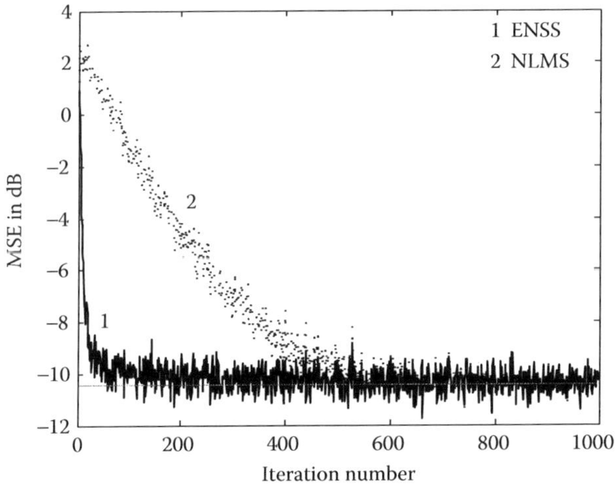

Figure 8.8.2 MSE learning curves of the ENSS and NLMS algorithms for white Gaussian input.

clearly provides faster speed of convergence than the NLMS algorithm. A MATLAB m file for the ENSS algorithm, which produces Figure 8.8.2, is shown below.

Book MATLAB function of the ENSS algorithm

```
% ENSS Algorithm in System Identification (See Figure
% 8.8.1)
% NN is the length of the error vector eL in the ENSS
% algorithm.
% mu: variable step-size of the ENSS algorithm.
% mu1: the dimensionless step-size of the ENSS algorithm.
%  N: length of the adaptive filter.
%  I: number of independent simulation runs used in the
%  plot of the learning curve.
%  LL: total number of iterations.
%   h: the impulse response of the unknown plant.
%  x is the input signal to both the adaptive filter and
%  the unknown system.
% dd is the output of the unknown system.
%   n: the internal noise of the unknown plant.
% J is the MSE.
clear all;
randn('state',0);
I=200; LL=1000; J=zeros(1,LL);
Jminn=zeros(1,LL);Jex=zeros(1,LL);
N=4; NN=10*N; h=[1 0.7 0.5 -0.2];
for i=1:I
     y=zeros(1,LL); w=zeros(1,N); e=zeros(1,LL);
     X=zeros(N,1); D=zeros(NN,1);
     x=sqrt(1)*randn(1,LL);
     denn=0; n=sqrt(0.09)*(randn(1,LL));
       for k=1:LL,
            dd=filter(h,1,x);
            X=[x(k); X(1:N-1)]; den=X'*X; y=w*X;
            e(k)=dd(k)+n(k)-y ;
            mu1=0.8; denn=denn+e(k)^2;mu=(mu1/(1+mu1*denn));
            w=w+mu*e(k)*X';
              J(k)=J(k)+(abs(e(k)))^2;
            Jminn(k)=Jminn(k)+n(k)^2;
     end;
end;
J=J/I; Jmin1=Jminn/I;Jmin=sum(Jmin1)/LL; Jex=J-Jmin;
Jinf=(1/200)*sum(J(LL-199:LL));JSSdB=10*log10(Jinf);
Jexinf=abs(Jinf-Jmin);JexinfSSdB=10*log10(Jexinf);
MM=Jexinf/Jmin;   Mpercent=MM*100;
[mu1,Jinf,Jmin,JSSdB,JexinfSSdB,Mpercent]
nn=0:LL-1;plot(nn,10*log10(abs((J))));
hold on;plot(nn,10*log10(Jmin));
```

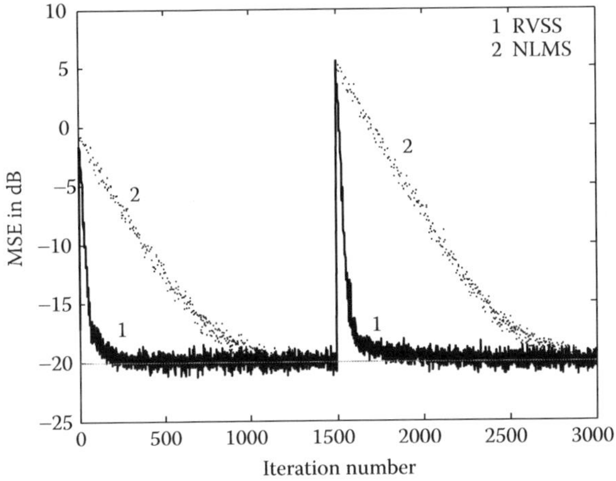

Figure 8.8.3 MSE learning curves of the RVSS and NLMS algorithms for an abrupt change in the plant parameters.

A comparison of the RVSS algorithm with the NLMS algorithms is demonstrated next for white Gaussian input in system identification setup. It is assumed that $\alpha = 0.5$, and $L = 10N$ in the RVSS algorithm. The length of the adaptive filter is assumed to be $N = 10$. The internal unknown system noise is white Gaussian with mean = 0 and variance σ_v^2 equals to 0.01. Simulation plots are obtained by ensemble averaging of 200 independent simulation runs. Figure 8.8.3 shows the learning curve of both algorithms for the case with an abrupt change in the impulse response of the plant, **h**. In particular, it is assumed that all the elements of **h** are multiplied by (−1) at iteration number 1500. Figure 8.8.4 shows the plot of the ensemble average trajectories of the fifth coefficient of the adaptive filter. The actual value of the corresponding unknown system coefficient to be identified is 0.5. The superiority of the RVSS algorithm is evident. A MATLAB m file for the RVSS algorithm with this assumed abrupt change in the plant parameters is shown below.

Book MATLAB function of the RVSS algorithm

```
% RVSS Algorithm in System Identification (See Figure
% 8.8.1)
% NN is the length of the error vector e_L in the RVSS
% algorithm
% mu: variable step-size of the RVSS algorithm.
% mu1: the dimensionless step-size of the RVSS algorithm.
% a= α in the algorithm.
%  N: length of the adaptive filter.
%  I: number of independent simulation runs used in the
%  plot of the learning curve.
%  LL: total number of iterations.
%  h: the impulse response of the unknown plant.
```

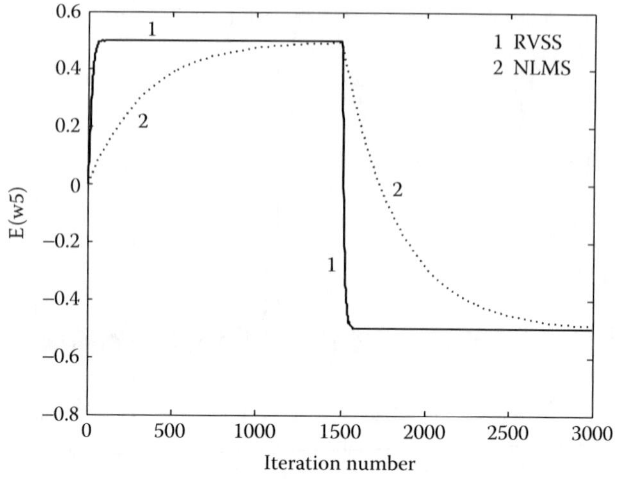

Figure 8.8.4 Ensemble averages of the 5th coefficient of the adaptive filter in the RVSS and NLMS algorithms for the case with an abrupt change in the plant parameters.

```
% x is the input signal to both the adaptive filter and
% the unknown system.
% dd is the output of the unknown system.
%  n: the internal noise of the unknown plant.
% J is the MSE and Jex is the excess MSE.
% Mpercent is the misadjustment percentage.
% JSSDB is the steady state MSE in dB.
clear all;
randn('state',0);
I=200; LL=3000; J=zeros(1,LL);
Jminn=zeros(1,LL);Jex=zeros(1,LL);
N=10; NN=10*N;
for i=1:I
     h=[0.1 0.2 0.3 0.4 0.5 0.4 0.3 0.2 0.1];
     y=zeros(1,LL); w=zeros(1,N); e=zeros(1,LL);
     X=zeros(N,1); D=zeros(NN,1);
     x=sqrt(1)*randn(1,LL);
     denn=0; n=sqrt(0.01)*(randn(1,LL));
        for k=1:LL,
           dd=filter(h,1,x);
           X=[x(k); X(1:N-1)]; den=X'*X; y=w*X;
           e(k)=dd(k)+n(k)-y ;
                          if k= =1500;
                               h=-[0.1 0.2 0.3 0.4 0.5...
     0.4 0.3 0.2 0.1];
                          end;
        mu1=0.07; D=[e(k); D(1:NN-1)];denx=(D'*D); a=0.5;
           denn=denn+e(k)^2; mu=(mu1*denx)/...
((a*denn+(1-a)*den));
```

```
            w=w+mu*e(k)*X';
            J(k)=J(k)+(abs(e(k)))^2;
            Jminn(k)=Jminn(k)+n(k)^2;
        end;
end;
J=J/I;  Jmin1=Jminn/I;Jmin=sum(Jmin1)/LL;  Jex=J-Jmin;
Jinf=(1/200)*sum(J(LL-199:LL));JSSdB=10*log10(Jinf);
Jexinf=abs(Jinf-Jmin);JexinfSSdB=10*log10(Jexinf);
MM=Jexinf/Jmin;    Mpercent=MM*100;
[mu1,Jinf,Jmin,JSSdB,JexinfSSdB,Mpercent]
nn=0:LL-1;plot(nn,10*log10(abs((J))));
hold on;plot(nn,10*log10(Jmin));
```

Finally, the performance of the EDNSS algorithm is compared with the NLMS algorithm in an adaptive noise canceller shown in Figure 8.8.5. The simulations are carried out using a male native speech saying "sound editing just gets easier and easier" sampled at a frequency of 11.025 kHz. The number of bits per sample is 8, and the total number of samples is 33000 or 3 sec of real time. The same value of step-size ($\mu = 0.1$) was used in both algorithms to achieve a compromise between small EMSE and high initial rate of convergence for a wide range of noise variances. In the EDNSS algorithm, we used $\alpha = 0.7$ and $L = 20N$. The order of the adaptive filter was assumed to be $N = 10$. Figure 8.8.6, from top to bottom, shows the original clean speech, corrupting noise with $\sigma_g^2 = 0.01$, speech corrupted by noise, and the recovered speech after noise cancellation using EDNSS algorithm. Listening tests show that the recovered speech is of a high quality and is very close to the original speech. Figure 8.8.7 compares the performance of the EDNSS algorithm with that of the NLMS for the case when $\sigma_g^2 = 0.01$. The figure shows plots of the EMSE in dB for that noise level of the two algorithms. While both algorithms have almost the same initial rate of convergence, the average EMSE in EDNSS is less than that of the NLMS by 10.6 dB. The values of EMSE were measured in both algorithms over all samples starting from sample number 2000, where the transient response has approximately ended. A MATLAB m file for the EDNSS algorithm in adaptive noise canceller for the above mentioned values of parameters is shown below.

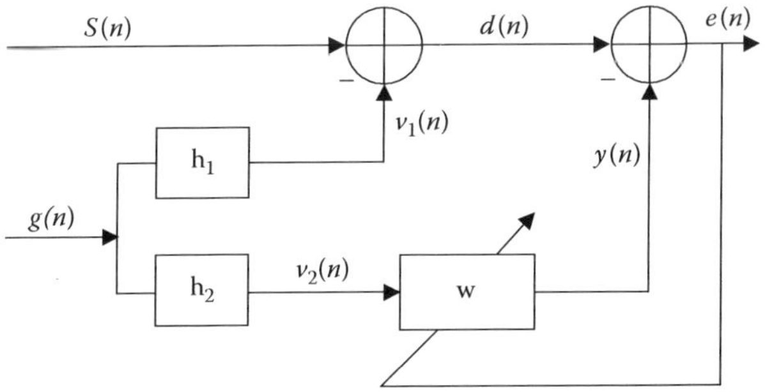

Figure 8.8.5 Adaptive noise canceller.

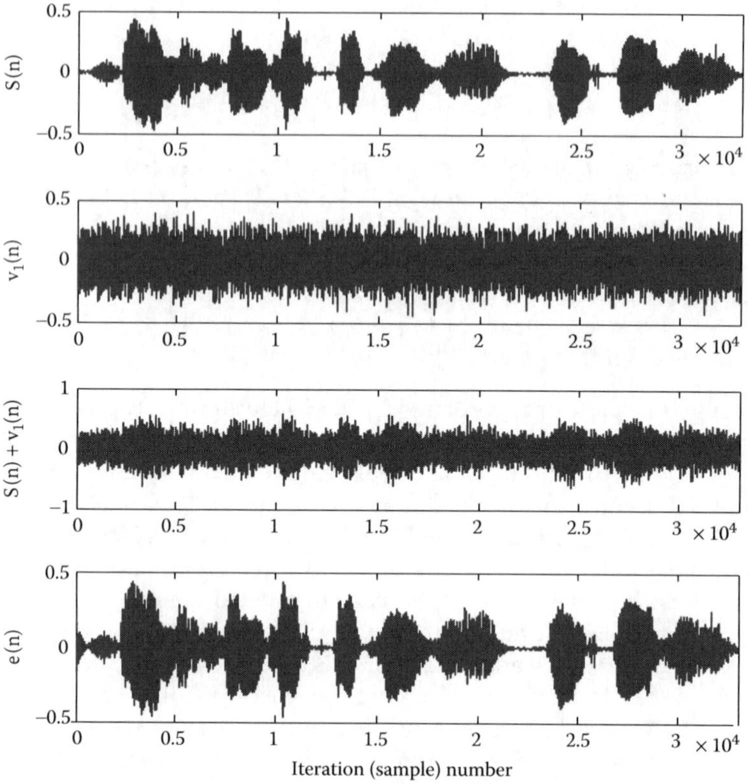

Figure 8.8.6 From top to bottom: Original clean speech S(*n*), noise that corrupts speech $v_1(n)$, corrupted speech $S(n) + v_1(n)$, recovered speech $e(n)$ using EDNSS algorithm ($\sigma_g^2 = 0.01$).

```
%EDNSS Algorithm in Adaptive Noise Canceller (see Figure
% 8.8.5).
% S: Speech signal vector and M is its length.
% LL=total number of iterations =M= total number of speech
% samples.
% N: length of the adaptive filter.
% fs: sampling frequency
% nbits: number of bits per second
% NN is the length of the error vector e_L in the EDNSS
% algorithm.
% h1 is the impulse responses of the autoregressive (AR)
% filter between the reference input and the primary
% microphone
% h2 is the impulse responses of the AR filter between
% the reference input and the adaptive filter w.
% JZ is the excess MSE smoothly averaged over N1 samples.
% Mu: Dimensionless step-size of the EDNSS algorithm.
```

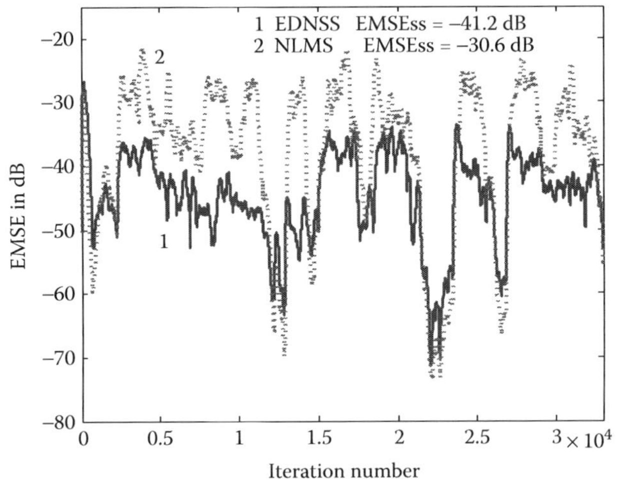

Figure 8.8.7 Excess mean-square error of the EDNSS and NLMS algorithms ($\sigma_g^2 = 0.01$).

```
% alpha: a parameter used in the EDNSS algorithm.
% e1 is the excess error.
[S,fs,nbits]=wavread('C:\sound_editing');
S=S(9500:42500);
S=S'; M=length(S); r1=(S*S')/M;
randn('state',0);rand('state',0);
h1=[1 -0.3 -.1];    h2=[1 -0.2];
N=10;   NN=20*N;     N1=200;    LL=M;
J=zeros(1,LL); J1=zeros(1,LL); JZ=zeros(1,LL);
g=sqrt(0.01)*randn(1,M);
v1=filter(1,h1,g); v2=filter(1,h2,g);
e=zeros(1,LL); y=zeros(1,LL);w=zeros(1,N);
V=zeros(N,1);E1=zeros(N1,1);D=zeros(NN,1);
d=S+v1;
for k=1:LL,
        V=[v2(k); V(1:N-1)];    den=V'*V;    y(k)=w*V;
      e(k)=d(k)-y(k);     e1(k)=e(k)-S(k);
     alpha1=0.7; Mu=0.1; D=[e(k); D(1:NN-1)];denx=(D'*D);
      w=w+(Mu/(alpha1*denx+(1-alpha1)*den))*e(k)*V';
   E1=[e1(k); E1(1:N1-1)]; JZ(k)=JZ(k)+((E1'*E1)/N1);
end
JZ=JZ/I;   F=2000;
JJZex=(1/(LL-F))*sum(JZ(F:LL-1));JJZexdB=10*log10(JJZex);
r11=(1/(LL-F))*S((F:LL-1))*S((F:LL-1))';
MM3=JJZex/r11;
```

```
[Mu,MM3*100,JJZexdB]
J11=J1(1:LL);nn=0:LL-1;
plot(nn,10*log10(abs((JZ))));
```

Table 8.8.1 shows a summary of the LMS-type algorithms presented in this chapter.

Table 8.8.1 Summary of the LMS Algorithms Presented in Chapter 8

$\mathbf{x}(n)=[x(n)x(n-1)\cdots x(n-M)]^T, \mathbf{w}(n)=[w_0(n)w_1(n)\cdots w_M(n)]^T, e(n)=d(n)-y(n)$

Algorithm	**Recursion**
1. LMS	$\mathbf{w}(n+1)=\mathbf{w}(n)+2\mu e(n)\mathbf{x}(n)$
2. LMS with complex data	$\mathbf{w}(n+1)=\mathbf{w}(n)+2\mu e^*(n)\mathbf{x}(n)$ $y(n)=\mathbf{w}^H(n)\mathbf{x}(n)(H=conjugate\ transpose)$
3. Sign LMS	$\mathbf{w}(n+1)=\mathbf{w}(n)+2\mu sign(e(n))\mathbf{x}(n)$
4. Sign-regressor LMS	$\mathbf{w}(n+1)=\mathbf{w}(n)+2\mu e(n)sign(\mathbf{x}(n))$
5. Sign-sign LMS	$\mathbf{w}(n+1)=\mathbf{w}(n)+2\mu sign(e(n))sign(\mathbf{x}(n))$
6. Normalized LM	$\mathbf{w}(n+1)=\mathbf{w}(n)+\frac{1}{\mathbf{x}^T(n)\mathbf{x}(n)}e(n)\mathbf{x}(n)$ $with\ \mu(n)=1/[2\mathbf{x}^T(n)\mathbf{x}(n)]$
9a. ε-Normalized LMS	$\mathbf{w}(n+1)=\mathbf{w}(n)+\frac{\bar{\mu}}{\varepsilon+\mathbf{x}^T(n)\mathbf{x}(n)}e(n)\mathbf{x}(n)$ $\bar{\mu}$ = step-size parameter ε = prevents division by very small number
9b. ε-Normalized LMS with complex data	$\mathbf{w}(n+1)=\mathbf{w}(n)+\frac{\bar{\mu}}{\varepsilon+\mathbf{x}^H(n)\mathbf{x}(n)}e^*(n)\mathbf{x}(n)$ H = conjugate transpose
10. Normalized LMS sign algorithm	$\mathbf{w}(n+1)=\mathbf{w}(n)+2\mu\frac{sign(e(n))\mathbf{x}(n)}{\varepsilon+\Vert\mathbf{x}(n)\Vert^2}$
11. Leaky LMS	$\mathbf{w}(n+1)=(1-2\mu\gamma)\mathbf{w}(n)+2\mu e(n)\mathbf{x}(n)$ 0 << <1
12. Constrained LMS	$\mathbf{w}(n+1)=\mathbf{w}'(n)+\frac{a-\mathbf{c}^T\mathbf{w}'(n)}{\mathbf{c}^T\mathbf{c}}\mathbf{c}$ $\mathbf{w}'(n)=\mathbf{w}(n)+2\mu e(n)\mathbf{x}(n)$ $\mathbf{c}$ = constant vector, a = constant
13. Self-correcting LMS	$y_{i+1}(n)=y_i(n)*w_{i+1}$ see also the m-file in the text
14. Transform domain LMS	see Sec. 8.7
15. Self-correcting adaptive filtering (SCAF)	see Sec. 8.6
16. ENSS Algorithm	see Sec. 8.8
17. RVSS Algorithm	see Sec. 8.8
18. EDNSS Algorithm	see Sec. 8.8

Problems

8.1.1 Show that the step-size of a sign algorithm is equivalent to $\mu'(n) = \mu/|e(n)|$. Discuss the value of $\mu'(n)$ for stability.

8.2.1 Minimize the posteriori error $e_{ps}(n) = d(n) - \mathbf{w}^{T}(n+1)\mathbf{x}(n)$ to obtain the normalized LMS algorithm.

8.4.1 To find the leaky LMS algorithm, apply the LMS gradient descent algorithm to minimize the error $J(n) = e^{2}(n) + \gamma\mathbf{w}^{T}(n)\mathbf{w}(n)$.

8.5.1 Verify (8.5.5).

8.7.1 Verify (8.7.10).

8.7.2 Verify (8.7.12).

8.7.3 Verify (8.7.15).

Hints-solutions-suggestions

8.1.1:
We write (8.1.1) in the form $\mathbf{w}(n+1) = \mathbf{w}(n) + 2\mu(e(n)/|e(n)|)\mathbf{x}(n)$. We observe that $\mu'(n) = 2\mu\frac{1}{|e(n)|}$ increases as $|e(n)|$ decreases. Therefore, μ must be very small for the logarithm to converge. If we choose very small μ, we automatically choose very small $\mu'(n)$, and thus, the convergence at the beginning is slow. But as $\mu'(n)$ increases the convergence becomes faster.

8.2.1:
Substituting the LMS recursion equation in $e_{ps}(n)$ we obtain $e_{ps}(n) = [1 - 2\mu(n)\mathbf{x}^{T}(n)\mathbf{x}(n)]e(n)$.

Hence, $$\frac{\partial e_{ps}^{2}(n)}{\partial\mu(n)} = 0 = -2\mathbf{x}^{T}(n)\mathbf{x}(n) - 2\mathbf{x}^{T}(n)\mathbf{x}(n) + 8\mu(n)\mathbf{x}^{T}(n)\mathbf{x}(n)\mathbf{x}^{T}(n)\mathbf{x}(n) \text{ or } \mu(n)$$

$$= 1/[2\mathbf{x}^{T}(n)\mathbf{x}(n)].$$

or

8.4.1:

$$J(n) = [d(n) - \mathbf{w}^{T}(n)\mathbf{x}(n)]^{2} + \gamma\mathbf{w}^{T}(n)\mathbf{w}(n) = d^{2}(n) - 2d(n)\mathbf{w}^{T}(n)\mathbf{x}(n)$$
$$+ \mathbf{w}^{T}(n)\mathbf{x}(n)\mathbf{w}^{T}(n)\mathbf{x}(n) + \gamma\mathbf{w}^{T}(n)\mathbf{w}(n) \quad (1)$$

(Note that $[\mathbf{w}^T(n)\mathbf{x}(n)]^2 = \mathbf{w}^T(n)\mathbf{x}(n)[\mathbf{w}^T(n)\mathbf{x}(n)]$ since $\mathbf{w}^T\mathbf{x}(n)$ is a number.) The ith component of (1) is $J(n) = d^2(n) - 2d(n)w_i(n)x_i(n) + [w_i(n)x_i(n)]^2 + \gamma w_i^2(n)$ (2). Therefore, $\partial J(n)/\partial w_i = 0 = -2d(n)x_i(n) + 2w_i(n)x_i(n) + 2\gamma w_i(n)$ (3). In matrix form it becomes $\nabla J = -e(n)\mathbf{x}(n) + \gamma\, \mathbf{w}(n)$ (4). Introducing (4) into the recursion equation of the gradient descent algorithm, we obtain

$$\mathbf{w}(n+1) = \mathbf{w}(n) - \mu\nabla J = \mathbf{w}(n) + \mu e(n)\mathbf{x}(n) - \mu\gamma\mathbf{w}(n) = (1-\mu\gamma)\mathbf{w}(n) + \mu e(n)\mathbf{x}(n).$$

8.5.1:

$$\begin{aligned} J &= E\{[d(n) - \mathbf{w}^T(n)\mathbf{x}(n)][d(n) - \mathbf{x}^T(n)\mathbf{w}(n)] + \lambda(\mathbf{c}^T\mathbf{w}(n) - a)\} \\ &= E\{d^2(n) - d(n)\mathbf{w}^T(n)\mathbf{x}(n) - d(n)\mathbf{x}^T(n)\mathbf{w}(n) + \mathbf{w}^T(n)\mathbf{x}(n)\mathbf{x}^T(n)\mathbf{w}(n) \\ &\quad + \lambda\mathbf{c}^T\mathbf{w}(n) - \lambda a\} \\ &= \sigma_d^2 - \mathbf{w}^T(n)E\{d(n)\mathbf{x}(n)\} - \mathbf{w}E\{d(n)\mathbf{x}(n)\} + \mathbf{w}^T(n)E\{\mathbf{x}(n)\mathbf{x}^T(n)\mathbf{w}(n) \\ &\quad + \lambda\mathbf{c}^T\mathbf{w}(n) - \lambda a \\ &= \sigma_d^2 - 2\mathbf{w}^T(n)\mathbf{p}_{\mathbf{dx}} + \mathbf{w}^T(n)\mathbf{R}_{\mathbf{x}}\mathbf{w}(n)\} + \lambda\mathbf{c}^T\mathbf{w}(n) - \lambda a \\ &= \sigma_d^2 - 2\mathbf{w}^T(n)\mathbf{p}_{\mathbf{dx}} + (\mathbf{w}(n) - \mathbf{w}^o)^T\mathbf{R}_{\mathbf{x}}(\mathbf{w}(n) - \mathbf{w}^o) + \mathbf{w}^{oT}\mathbf{R}_{\mathbf{x}}\,\mathbf{w}(n) + \mathbf{w}^T(n)\mathbf{R}_{\mathbf{x}}\,\mathbf{w}^o \\ &\quad + \mathbf{w}^{oT}\mathbf{R}_{\mathbf{x}}\mathbf{w}^o + \lambda[\mathbf{c}^T(\mathbf{w}(n) - \mathbf{w}^o) - (a - \mathbf{c}^T\mathbf{w}^o)] \\ &= \sigma_d^2 - \mathbf{w}^{oT}\mathbf{p}_{\mathbf{dx}} + \xi^T\mathbf{R}_{\mathbf{x}}\xi + \lambda(\mathbf{c}^T\xi - a') \end{aligned}$$

where $\mathbf{R}_{\mathbf{x}}\mathbf{w}^o = \mathbf{p}_{\mathbf{dx}}$, $(\mathbf{w}^{oT}\mathbf{R}_{\mathbf{x}}\mathbf{w}(n))^T = \mathbf{w}^T(n)\mathbf{R}_{\mathbf{x}}\mathbf{w}^o$ ($\mathbf{R}_{\mathbf{x}}$ = symmetric).

8.7.1:
The kth value is

$$\begin{aligned} W_{i+1,k} &= W_{i,k} + 2\mu X_i^* E_i = W_{i,k} + 2\mu X_i^*[D_i(k) - Y_i(k)] = W_{i,k} + 2\mu X_i^*(k)[D_i(k) \\ &\quad - W_{i,k}X_i(k)] \\ &= W_{i,k} + 2\mu X_i^*(k)D_i(k) - 2\mu W_{i,k}|X_i(k)|^2 = (1 - 2\mu|X_i(k)|^2)W_{i,k} \\ &\quad + 2\mu D_i(k)X_i^*(k) \end{aligned}$$

8.7.2:
The z-transform and the ensemble are linear operations and can be interchanged. Therefore, the other z-transform is

$$zW_k(z) - zW_{0,k} = (1 - 2\mu E\{|X_i(k)|^2\})W_k(z) + [2\mu E\{D_i(k)X_i^*(k)\}/(1 - z^{-1})] \text{ (1)},$$

where $W_{0,k} = 0$ since it was assumed that the initial conditions have zero values, and $W_k(z) = \sum_{i=0}^{\infty} E\{W_{i,k}\}z^{-i}$. Multiplying (1) by z^{-1} and $(z-1)$, and applying the final value theorem (see Table 2.3.1), we obtain

$$E\{W_k^{\infty}\} = \lim_{z\to 1}\{z-1)W_k(z)\} = \lim_{z\to 1}\frac{2\mu E\{D_i(k)X_i^*(k)\}}{1-(1-2\mu E\{|X_i(k)|^2\})z^{-1}} = \frac{E\{D_i(k)X_i^*(k)\}}{E\{|X_i(k)|^2\}} \quad (2)$$

8.7.3:

$$E_{i+1}(k) = E\{W_{i+1,k}\} - E\{W_k^{\infty}\} = (1-2\mu E\{|X_i(k)|^2\})E\{W_{i,k}\} + 2\mu E\{D_i(k)X_i^*(k)\}$$

$$-\frac{E\{D_i(k)X_i^*(k)\}}{E\{|X_i(k)|^2\}} = (1-2\mu E\{|X_i(k)|^2\})E\{W_{i,k}\} - (1-2\mu E\{|X_i(k)|^2\})\frac{E\{D_i(k)X_i^*(k)\}}{E\{|X_i(k)|^2\}}$$

$$= (1-2\mu E\{|X_i(k)|^2\})E_i(k)$$

chapter 9

Least squares and recursive least-squares signal processing

9.1 Introduction to least squares

The Wiener and adaptive filters belong to the statistical framework since the signal statistics are being invoked and it is required that *a priori* knowledge exists of the second-order moments. On the other hand, the method of least squares belongs to the deterministic frame. In addition, this method requires that both the input signal and the desired one be measured. There are several important cases that such restrictions of signal measurements can be applied, such as modeling applications, linear predictive coding, and communications, where the desired signal is taken to be the training set.

9.2 Least-square formulation

We consider a linear adaptive filter with coefficients at time n

$\mathbf{w}(n) = [w_1(n)\ w_2(n) \cdots w_M(n)]^{\mathrm{T}}$, a measured real-valued input vector

$\mathbf{x}(n) = [x_1(n)\ x_2(n)\ \cdots\ x_M(n)]^{\mathrm{T}}$, and a measured desired response $d(n)$.

Note that no structure has been specified for the input vector $\mathbf{x}(n)$, and therefore, it can be considered as the successive samples of a particular process or as a snapshot of M detectors as shown in Figure 9.2.1. Hence, the problem is to estimate the desired response $d(n)$ using the linear combination

$$y(n) = \mathbf{w}^{\mathrm{T}}(n)\mathbf{x}(n) = \sum_{k=1}^{M} w_k(n) x_k(n) \quad n = 1, 2, \cdots, N \tag{9.2.1}$$

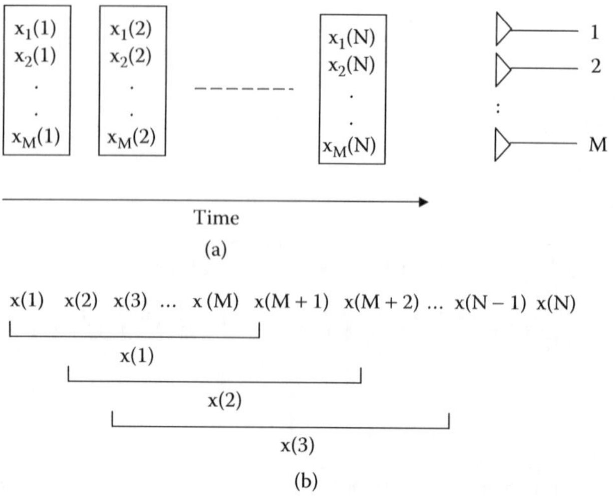

Figure 9.2.1 (a) Multisensor application. (b) Single sensor application.

The above equation can be represented by a linear combiner as shown in Figure 9.2.2. The estimation error is defined by the relation

$$e(n) = d(n) - y(n) = d(n) - \mathbf{w}^T(n)\mathbf{x}(n) \tag{9.2.2}$$

The coefficients of the adaptive filter are found by minimizing the sum of the squares of the error (least squares)

$$J \equiv E = \sum_{n=1}^{N} g(n)e^2(n) \tag{9.2.3}$$

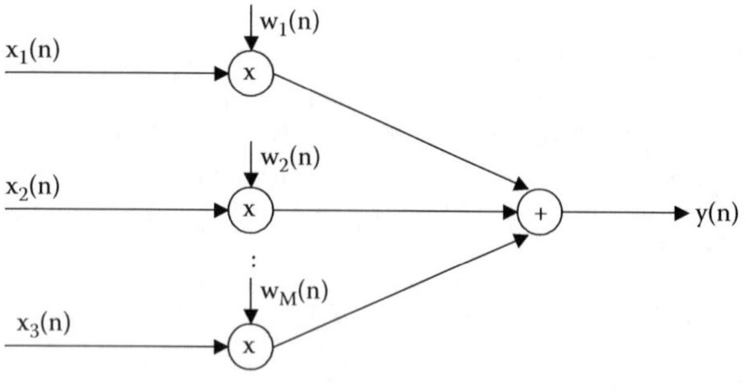

Figure 9.2.2 Linear estimator (*M*-parameter system).

where $g(n)$ is a weighting function. Therefore, in the method of least squares the filter coefficients are optimized by using all the observations from the time the filter begins until the present time and minimizing the sum of the squared values of the error samples that are equal to the measured desired signal and the output signal of the filter. The minimization is valid when the filter coefficient vector $\mathbf{w}(n)$ is kept constant, $\mathbf{w}$, over the measurement time interval $1 \le n \le N$. In statistics, the least-squares estimation is known as *regression*, $e(n)$ are known as *signals*, and $\mathbf{w}$ is the *regression vector.*

We next define the matrix of the observed input samples as

$$\mathbf{X}^{\mathrm{T}} = \begin{bmatrix} x_1(1) & x_1(2) & \cdots & x_1(N) \\ x_2(1) & x_2(2) & \cdots & x_2(N) \\ \vdots & \vdots & & \vdots \\ x_M(1) & x_M(2) & & x_M(N) \end{bmatrix} \quad \underset{\textit{snapshots}}{\downarrow} \quad \rightarrow \quad \textit{data records} \quad (M \times N) \tag{9.2.4}$$

where we assume that $N > M$. This defines an over-determined least-squares problem.

For the case in which we have one dimensional input signal, as shown in Figure 9.2.1b, the data matrix takes the form

$$\mathbf{X}^{\mathrm{T}} = \begin{bmatrix} x(M) & x(M+1) & \cdots & x(N) \\ x(M-1) & x(M) & \cdots & x(N-1) \\ x(M-2) & x(M-1) & \cdots & x(N-2) \\ \vdots & \vdots & & \vdots \\ x(1) & x(2) & \cdots & x(N-M+1) \end{bmatrix} \tag{9.2.5}$$

The output $\mathbf{y}$, the error $\mathbf{e}$, and the data vectors $\mathbf{x}_k$, are:

$$\mathbf{y} = \mathbf{X}\mathbf{w} \tag{9.2.6}$$

$$\mathbf{e} = \mathbf{d} - \mathbf{y} \tag{9.2.7}$$

where

$$\mathbf{y} = [y(1)\ \ y(2)\ \cdots\ y(N)]^{\mathrm{T}} = \text{filter output vector } (\mathrm{N} \times 1) \tag{9.2.8}$$

$$\mathbf{d} = [d(1)\ \ d(2)\ \cdots\ d(N)]^{\mathrm{T}} \tag{9.2.9}$$

$$\mathbf{e} = [e(1)\ \ e(2)\ \cdots\ e(N)]^{\mathrm{T}} \equiv \text{error vector } (\mathrm{N} \times 1) \tag{9.2.10}$$

$$\mathbf{x} = [x_1(n)\ \ x_2(n)\ \cdots\ x_N(n)]^{\mathrm{T}} \equiv \text{snapshot } (\mathrm{N} \times 1) \tag{9.2.11}$$

$$\mathbf{x}_k = [x_k(1)\ \ x_k(2)\ \cdots\ x_k(N)]^{\mathrm{T}} \equiv \text{data vector, } \ \mathrm{k} = 1, 2, \cdots, \mathrm{M} \tag{9.2.12}$$

$$\mathbf{w} = [w_1\ w_2\ \cdots\ w_M]^{\mathrm{T}} \equiv \text{filter coefficients, } (\mathrm{M} \times 1) \tag{9.2.13}$$

$$\mathbf{X} = [\mathbf{x}_1\, \mathbf{x}_2 \cdots \mathbf{x}_M]^{\mathrm{T}} \equiv (\mathrm{N} \times \mathrm{M}) \tag{9.2.14}$$

In addition, with $g(n) = 1$ for all n, (9.2.3) takes the form

$$\begin{aligned} J &= \mathbf{e}^{\mathrm{T}}\mathbf{e} = (\mathbf{d} - \mathbf{y})^{\mathrm{T}}(\mathbf{d} - \mathbf{y}) = (\mathbf{d} - \mathbf{X}\mathbf{w})^{\mathrm{T}}(\mathbf{d} - \mathbf{X}\mathbf{w}) \\ &= \mathbf{d}^{\mathrm{T}}\mathbf{d} - \mathbf{w}^{\mathrm{T}}\mathbf{X}^{\mathrm{T}}\mathbf{d} - \mathbf{d}^{\mathrm{T}}\mathbf{X}\mathbf{w} + \mathbf{w}^{\mathrm{T}}\mathbf{X}^{\mathrm{T}}\mathbf{X}\mathbf{w} \\ &= E_d - \mathbf{w}^{\mathrm{T}}\mathbf{p} - \mathbf{p}^{\mathrm{T}}\mathbf{w} + \mathbf{w}^{\mathrm{T}}\mathbf{R}\mathbf{w} = E_d - 2\mathbf{p}^{\mathrm{T}}\mathbf{w} + \mathbf{w}^{\mathrm{T}}\mathbf{R}\mathbf{w} \end{aligned} \tag{9.2.15}$$

where

$$E_d = \mathbf{d}^{\mathrm{T}}\mathbf{d} = \sum_{n=1}^{N} d(n)d(n) \tag{9.2.16}$$

$$\mathbf{R} = \mathbf{X}^{\mathrm{T}}\mathbf{X} = \sum_{n=1}^{N} \mathbf{x}(n)\mathbf{x}^{\mathrm{T}}(n) \quad (M \times M) \tag{9.2.17}$$

$$\mathbf{p} = \mathbf{X}^{\mathrm{T}}\mathbf{d} = \sum_{n=1}^{N} \mathbf{x}(n)d(n) \quad (M \times 1) \tag{9.2.18}$$

$$\mathbf{y} = \mathbf{X}\mathbf{w} = \sum_{\mathrm{k}=1}^{\mathrm{M}} w_k \mathbf{x}_k \quad (N \times 1) \tag{9.2.19}$$

The matrix **R** becomes time averaged if it is divided by N. In statistics, the scaled form of **R** is known as the *sample correlation matrix.*

Setting the gradient of J with respect to the vector coefficients **w** equal to zero, we obtain (see Problem 9.2.1)

$$\mathbf{R}\hat{\mathbf{w}} = \mathbf{p}; \quad \mathbf{p}^T = \hat{\mathbf{w}}^T \mathbf{R}^T = \hat{\mathbf{w}}^T \mathbf{R} \quad (\mathbf{R} \text{ is symmetric}) \tag{9.2.20}$$

or

$$\hat{\mathbf{w}} = \mathbf{R}^{-1}\mathbf{p} \tag{9.2.21}$$

Therefore, the minimum sum of squared errors is given by

$$J_{min} = \mathbf{d}^T\mathbf{d} - 2\mathbf{p}^T\mathbf{R}^{-1}\mathbf{p} + \mathbf{w}^T\mathbf{R}\mathbf{R}^{-1}\mathbf{p} = E_d - \mathbf{p}^T\mathbf{R}^{-1}\mathbf{p} = E_d - \mathbf{p}^T\hat{\mathbf{w}} \tag{9.2.22}$$

since **R** is symmetric. For the solution given in (9.2.21) see Problem 9.2.3 and Problem 9.2.4.

Example 9.2.1: Let the desired response be $d = [1 \;\; 1 \;\; 1 \;\; 1]$, and the two measured signals be $x_1 = [0.7 \;\; 1.4 \;\; 0.4 \;\; 1.3]^T$, $x_2 = [1.2 \;\; 0.6 \;\; 0.5 \;\; 1.1]^T$. Then we obtain

$$\mathbf{R} = \mathbf{X}^T\mathbf{X} = \begin{bmatrix} 0.7 & 1.4 & 0.4 & 1.3 \\ 1.2 & 0.6 & 0.5 & 1.1 \end{bmatrix} \begin{bmatrix} 0.7 & 1.2 \\ 1.4 & 0.6 \\ 0.4 & 0.5 \\ 1.3 & 1.1 \end{bmatrix} = \begin{bmatrix} 4.30 & 3.31 \\ 3.31 & 3.26 \end{bmatrix}$$

$$\mathbf{p} = \mathbf{X}^T\mathbf{d} = \begin{bmatrix} 3.8 \\ 3.4 \end{bmatrix}, \quad \hat{\mathbf{w}} = \mathbf{R}^{-1}\mathbf{p} = \begin{bmatrix} 0.3704 \\ 0.6669 \end{bmatrix}, \quad J_{min} = 0.3252$$

$$\hat{\mathbf{y}} = \mathbf{X}\,\hat{\mathbf{w}} = [1.0595 \quad 0.9187 \quad 0.4816 \quad 1.2150]$$

The least-squares technique is a mathematical procedure that enables us to achieve a best fit of a model to experimental data. In the sense of the M-parameter linear system, shown in Figure 9.2.3, (9.2.1) is written in the form

$$y(n) = w_1 x_1(n) + w_2 x_2(n) + \cdots + w_M x_M(n) \quad n = 1, 2, \cdots, N \tag{9.2.23}$$

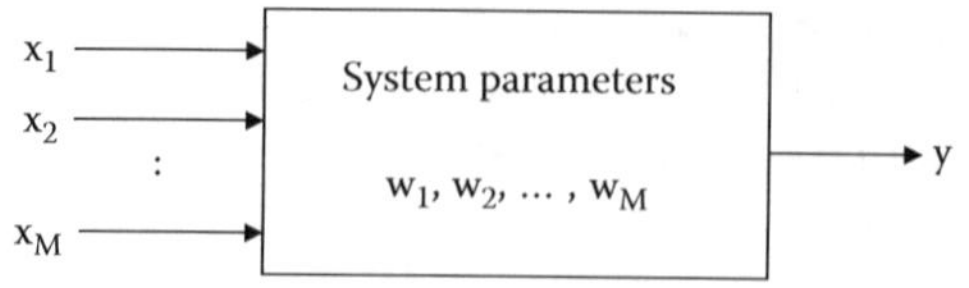

Figure 9.2.3 An *M*-parameter linear system.

The above equation takes the following matrix form

$$\mathbf{y} = \mathbf{X}\mathbf{w} \tag{9.2.24}$$

To estimate the M parameters w_i, it is necessary that $N \geq M$. If $N = M$, then we can uniquely solve for $\mathbf{w}$ to find

$$\hat{\mathbf{w}} = \mathbf{X}^{-1}\mathbf{y} \tag{9.2.25}$$

provided $\mathbf{X}^{-1}$ exists. $\hat{\mathbf{w}}$ is the estimate of $\mathbf{w}$. Using the least-error-squares we can determine $\mathbf{w}$ provided that $N > M$.

Let us define an error vector $\mathbf{e} = [e_1 \; e_2 \cdots e_N]^{\mathrm{T}}$ as follows:

$$\mathbf{e} = \mathbf{y} - \mathbf{X}\mathbf{w} \tag{9.2.26}$$

Next we choose $\hat{\mathbf{w}}$ in such a way that the criterion

$$J = \sum_{i=1}^{N} e_i^2 = \mathbf{e}^{\mathrm{T}}\mathbf{e} \tag{9.2.27}$$

is minimized. To proceed we write

$$\begin{aligned} J &= (\mathbf{y} - \mathbf{X}\mathbf{w})^{\mathrm{T}}(\mathbf{y} - \mathbf{X}\mathbf{w}) \\ &= \mathbf{y}^{\mathrm{T}}\mathbf{y} - \mathbf{w}^{\mathrm{T}}\mathbf{X}^{\mathrm{T}}\mathbf{y} - \mathbf{y}\mathbf{X}\mathbf{w} + \mathbf{w}^{\mathrm{T}}\mathbf{X}^{\mathrm{T}}\mathbf{X}\mathbf{w} \end{aligned} \tag{9.2.28}$$

Differentiating J with respect to $\mathbf{w}$ and equating the result to zero for determining the conditions on the estimate $\hat{\mathbf{w}}$ that minimizes J. Hence,

$$\left.\frac{\partial J}{\partial \mathbf{w}}\right|_{\mathbf{w}=\hat{\mathbf{w}}} = -2\mathbf{X}^{\mathrm{T}}\mathbf{y} + 2\mathbf{X}^{\mathrm{T}}\mathbf{X}\hat{\mathbf{w}} = \mathbf{0} \tag{9.2.29}$$

$$\mathbf{X}^{\mathrm{T}}\mathbf{X}\hat{\mathbf{w}} = \mathbf{X}^{\mathrm{T}}\mathbf{y} \tag{9.2.30}$$

from which we obtain

$$\hat{\mathbf{w}} = (\mathbf{X}^T\mathbf{X})^{-1}\mathbf{X}^T\mathbf{y} \tag{9.2.31}$$

The above is known as the *least-squares estimator* (LSE) of $\mathbf{w}$. (9.2.30) is known as the *normal equations.*

If we weight differently each error term, then the weighted error criterion becomes

$$J_G = \mathbf{e}^T\mathbf{G}\mathbf{e} = (\mathbf{y} - \mathbf{X}\mathbf{w})^T\mathbf{G}(\mathbf{y} - \mathbf{X}\mathbf{w}) \tag{9.2.32}$$

The weighting matrix $\mathbf{G}$ is restricted to be symmetric positive definite matrix. Minimizing J_G with respect to $\mathbf{w}$ results in the following weighted least-squares estimator (WLSE) $\hat{\mathbf{w}}$:

$$\hat{\mathbf{w}} = (\mathbf{X}^T\mathbf{G}\mathbf{X})^{-1}\mathbf{X}^T\mathbf{G}\mathbf{y} \tag{9.2.33}$$

If $\mathbf{G} = \mathbf{I}$ then $\hat{\mathbf{w}} = \hat{\mathbf{w}}\mathbf{G}$.

Statistical properties of least-squares estimators

We rewrite (9.2.26) in the form ($\mathbf{X}$ = deterministic matrix)

$$\mathbf{y} = \mathbf{X}\mathbf{w} + \mathbf{e} \tag{9.2.34}$$

and assume that $\mathbf{e}$ is a stationary random vector with zero mean value, $E[\mathbf{e}] = \mathbf{0}$. Furthermore, $\mathbf{e}$ is assumed to be uncorrelated with $\mathbf{y}$ and $\mathbf{X}$. Therefore, on the given statistical properties of $\mathbf{e}$, we wish to know just how good, or how accurate, the estimates of the parameters are.

Substituting (9.2.34) in (9.2.31) and taking the ensemble average we obtain

$$\begin{aligned} E\{\hat{\mathbf{w}}\} &= E\{\mathbf{w} + (\mathbf{X}^T\mathbf{X})^{-1}\mathbf{X}^T\mathbf{e}\} = E\{\mathbf{w}\} + E\{\mathbf{X}^T\mathbf{X})^{-1}\mathbf{X}\}E\{\mathbf{e}\} \\ &= \mathbf{w} \qquad (E\{\mathbf{e}\} = \mathbf{0}) \end{aligned} \tag{9.2.35}$$

which indicates that $\hat{\mathbf{w}}$ is *unbiased.*

The covariance matrix corresponding to the estimate error $\hat{\mathbf{w}} - \mathbf{w}$ is

$$\begin{aligned} \mathbf{C}_w &\equiv E\{(\hat{\mathbf{w}} - \mathbf{w})(\hat{\mathbf{w}} - \mathbf{w})^T\} = E\{[(\mathbf{X}^T\mathbf{X})^{-1}\mathbf{X}^T\mathbf{y} - \mathbf{w}](\hat{\mathbf{w}} \quad \mathbf{w})^T\} \\ &= E\{[\mathbf{X}^T\mathbf{X})^{-1}\mathbf{X}^T(\mathbf{X}\mathbf{w} + \mathbf{e}) - \mathbf{w}](\hat{\mathbf{w}} - \mathbf{w})^T\} \\ &= E\{[(\mathbf{X}^T\mathbf{X})^{-1}(\mathbf{X}^T\mathbf{X})\mathbf{w} + (\mathbf{X}^T\mathbf{X})^{-1}\mathbf{e} - \mathbf{w}][\hat{\mathbf{w}} - \mathbf{w}]^T\} \\ &= E\{[(\mathbf{X}^T\mathbf{X})^{-1}\mathbf{X}^T\mathbf{e}][(\mathbf{X}^T\mathbf{X})^{-1}\mathbf{X}^T\mathbf{e}]^T\} \\ &= (\mathbf{X}^T\mathbf{X})^{-1}\mathbf{X}^T E\{\mathbf{e}\mathbf{e}^T\}\mathbf{X}(\mathbf{X}^T\mathbf{X})^{-1} \\ &= (\mathbf{X}^T\mathbf{X})^{-1}\mathbf{X}^T\mathbf{R}_e\mathbf{X}(\mathbf{X}^T\mathbf{X})^{-1} \quad (\mathbf{R}_e \text{ is the error correlation marix}) \end{aligned} \tag{9.2.36}$$

If the noise samples $e(i)$ for $i = 1, 2, 3\ldots$ are normal, identical distributed with zero mean and variance σ^2 $((\mathbf{e} = N(\mathbf{0}, \sigma\mathbf{I}))$, then

$$\mathbf{R}_{\mathbf{e}} = E\{\mathbf{e}\mathbf{e}^{\mathrm{T}}\} = \sigma^2\mathbf{I} \tag{9.2.37}$$

and, hence,

$$\mathbf{C}_{\mathbf{w}} = \sigma^2(\mathbf{X}^{\mathrm{T}}\mathbf{X})^{-1} \tag{9.2.38}$$

Using (9.2.34), and taking into consideration that $\mathbf{e}$ is a Gaussian random vector, then the natural logarithm of its probability density is given by

$$\begin{aligned}\ln p(\mathbf{e};\mathbf{w}) &= \ln\left[\frac{1}{(2\pi)^{N/2}}\frac{1}{|\mathbf{C}_{\mathbf{e}}|}\exp\left[-\frac{1}{2}(\mathbf{y}-\mathbf{X}\mathbf{w})^{\mathrm{T}}\mathbf{C}_{\mathbf{e}}^{-1}(\mathbf{y}-\mathbf{X}\mathbf{w})\right]\right] \\ &= -\ln(2\pi\sigma^2)^{N/2} - \frac{1}{2\sigma^2}(\mathbf{y}-\mathbf{X}\mathbf{w})^{\mathrm{T}}(\mathbf{y}-\mathbf{X}\mathbf{w})\end{aligned} \tag{9.2.39}$$

since $\mathbf{C}_{\mathbf{e}} = \sigma^2\mathbf{I}$ and $|\mathbf{C}_{\mathbf{e}}|$ implies the determinant of $\mathbf{C}_{\mathbf{e}}$. Next, we differentiate (9.2.39) with respect to the parameter $\mathbf{w}$. Hence, we find

$$\frac{\partial \ln(\mathbf{e};\mathbf{w})}{\partial \mathbf{w}} = -\frac{1}{2\sigma^2}\frac{\partial}{\partial \mathbf{w}}[\mathbf{y}^{\mathrm{T}}\mathbf{y} - 2\mathbf{y}^{\mathrm{T}}\mathbf{X}\mathbf{w} + \mathbf{w}^{\mathrm{T}}\mathbf{X}^{\mathrm{T}}\mathbf{X}\mathbf{w}] \tag{9.2.40}$$

since $\mathbf{y}^{\mathrm{T}}\mathbf{X}\mathbf{w} = \mathbf{w}^{\mathrm{T}}\mathbf{X}^{\mathrm{T}}\mathbf{y}$ = scalar. Using the identities below (see Appendix A)

$$\frac{\partial \mathbf{b}^{\mathrm{T}}\mathbf{w}}{\partial \mathbf{w}} = \mathbf{b}, \qquad \frac{\partial \mathbf{w}^{\mathrm{T}}\mathbf{A}\mathbf{w}}{\partial \mathbf{w}} = 2\mathbf{A}\mathbf{w} \quad (\mathbf{A}\text{ is symmetric}) \tag{9.2.41}$$

(9.2.40) becomes

$$\frac{\partial \ln p(\mathbf{e};\mathbf{w})}{\partial \mathbf{w}} = \frac{1}{\sigma^2}[\mathbf{X}^{\mathrm{T}}\mathbf{y} - \mathbf{X}^{\mathrm{T}}\mathbf{X}\mathbf{w}] \tag{9.2.42}$$

Assuming that $\mathbf{X}^{\mathrm{T}}\mathbf{X}$ is invertible, then

$$\frac{\partial \ln p(\mathbf{e};\mathbf{p})}{\partial \mathbf{w}} = \frac{\mathbf{X}^{\mathrm{T}}\mathbf{X}}{\sigma^2}[(\mathbf{X}^{\mathrm{T}}\mathbf{X})^{-1}\mathbf{X}^{\mathrm{T}}\mathbf{y} - \mathbf{w}] = \mathbf{I}(\mathbf{w})[\mathbf{g}(\mathbf{w}) - \mathbf{w}] \tag{9.2.43}$$

From the Crame–Rao lower bound (CRLB) theorem, $\hat{\mathbf{w}}$ is the minimum variance unbiased (MVU) estimator, since we have found that

$$\hat{\mathbf{w}} = (\mathbf{X}^{\mathrm{T}}\mathbf{X})^{-1}\mathbf{X}^{\mathrm{T}}\mathbf{y} \equiv \mathbf{g}(\mathbf{w}) \tag{9.2.44}$$

and (9.2.43) takes the form

$$\frac{\partial \ln p(\mathbf{e};\mathbf{w})}{\partial \mathbf{w}} = \frac{\mathbf{X}^{\mathrm{T}}\mathbf{X}}{\sigma^2}(\hat{\mathbf{w}} - \mathbf{w}) \tag{9.2.45}$$

The matrix

$$\mathbf{I}(\mathbf{w}) = \mathbf{X}^{\mathrm{T}}\mathbf{X}/\sigma^2 \tag{9.2.46}$$

is known as the *Fisher information matrix*. The Fisher matrix is defined by the relation

$$[\mathbf{I}(\mathbf{w})]_{ij} = E\left\{\frac{\partial^2 \ln p(\mathbf{e};\mathbf{w})}{\partial w_i \partial w_j}\right\} \tag{9.2.47}$$

in the CRLB theorem, and thus, the parameters are shown explicitly. Comparing (9.2.38) and (9.2.46), the MVU estimator of $\mathbf{w}$ is given by (9.2.44) and its covariance matrix is

$$\mathbf{C}_{\mathbf{w}} = \mathbf{I}^{-1}(\mathbf{w}) = \sigma^2(\mathbf{X}^{\mathrm{T}}\mathbf{X})^{-1} \tag{9.2.48}$$

The MVU estimator of the linear model (9.2.34) is *efficient* since it attains the CRLB or, in other words, the covariance matrix is equal to the inverse of the Fisher information matrix.

Let us rewrite the error covariance matrix in the form

$$\mathbf{C}_{\mathbf{w}} = \sigma^2(\mathbf{X}^{\mathrm{T}}\mathbf{X})^{-1} = \frac{\sigma^2}{N}\left(\frac{1}{N}\mathbf{X}^{\mathrm{T}}\mathbf{X}\right)^{-1} \tag{9.2.49}$$

where N is the number of equations in the vector equation (9.2.34). Let $\lim_{N\to\infty}[(1/N)\mathbf{X}^{\mathrm{T}}\mathbf{X}]^{-1} = \mathbf{A}$ where $\mathbf{A}$ is a rectangular constant matrix. Then

$$\lim_{N\to\infty}\mathbf{C}_{\mathbf{w}} = \lim_{N\to\infty}\frac{\sigma^2\mathbf{A}}{N} = \mathbf{0} \tag{9.2.50}$$

Since the covariance is zero as N goes to infinity implies that $\hat{\mathbf{w}} = \mathbf{w}$. The above convergence property defines $\hat{\mathbf{w}}$ as a *consistent estimator.*

The above development shows if a system is modeled as linear in the presence of white Gaussian noise, the LSE approach provides estimators that are unbiased and consistent.

9.3 Least-squares approach

Using the least-squares (LS) approach we try to minimize the squared difference between the given data (or desired data) $d(n)$ and the output signal of a LTI system. The signal $y(n)$ is generated by some system, which in turn depends upon its unknown parameters w_i's. The LSE of w_i's chooses the values that make y's closest to the given data. The measure of closeness is defined by the LSE (see also (9.2.15)). For the one-coefficient system model, we have

$$J(w) = \sum_{n=1}^{N} (d(n) - y(n))^2 \tag{9.3.1}$$

and the dependence of J on $\mathbf{w}$ is via $y(n)$. The value of $\mathbf{w}$ that minimizes the cost function $J(\mathbf{w})$ is the LSE. It is apparent that the performance of LSE will depend upon the statistical properties of the corrupting noise to the signal as well as any system modeling error.

Example 9.3.1: Let us assume that the signal is $y(n) = acos(_o n)$, where ω_o is known and the amplitude a must be determined. Hence, the LSE minimizes the cost function

$$J(a) = \sum_{n=1}^{N} (d(n) - a\cos\omega_o n)^n \tag{9.3.2}$$

Therefore, we obtain

$$\frac{\partial J(a)}{\partial a} = \sum_{n=1}^{N} (-)2\cos\omega_o n(d(n) - a\cos\omega_o n) = 0$$

$$\hat{a} = \frac{\sum_{n=1}^{N} d(n)\cos\omega_o n}{\sum_{n=1}^{N} \cos\omega_o n}$$

Let us assume that the output of a system is linear, and it is given by the relation $y(n) = x(n)w$, where $x(n)$ is a known sequence. Hence, the LSE criterion becomes

$$J(w) = \sum_{n=1}^{N} (d(n) - x(n)w)^2 \tag{9.3.3}$$

The estimate value of **w** is

$$\hat{\mathbf{w}} = \frac{\sum_{n=1}^{N} d(n)x(n)}{\sum_{n=1}^{N} x^2(n)} \tag{9.3.4}$$

and the minimum LS error is given by (Problem 9.3.2)

$$J_{\min} = J(\hat{w}) = \sum_{n=1}^{N} d^2(n) - \hat{w}\sum_{n=1}^{N} d(n)x(n) = \sum_{n=1}^{N} d^2(n) - \frac{\left(\sum_{n=1}^{N} d(n)x(n)\right)^2}{\sum_{n=1}^{N} x^2(n)} \tag{9.3.5}$$

Example 9.3.2: Consider the experimental data shown in Figure 9.3.1. It is recommended that a linear model, $y(n) = a + bn$, for the data be used. Using the LSE approach, we find the cost function

$$J(\mathbf{w}) = \sum_{n=1}^{N} (d(n) - a - bn)^2 = (\mathbf{d} - \mathbf{X}\mathbf{w})^{\mathrm{T}}(\mathbf{d} - \mathbf{X}\mathbf{w}) \tag{9.3.6}$$

where

$$\mathbf{w} = \begin{bmatrix} a \\ b \end{bmatrix}; \quad \mathbf{X} = \begin{bmatrix} 1 & 1 \\ 1 & 2 \\ \vdots & \vdots \\ 1 & N \end{bmatrix} \tag{9.3.7}$$

From (9.2.31), the estimate value of **w** is

$$\hat{\mathbf{w}} = (\mathbf{X}^{\mathrm{T}}\mathbf{X})^{-1}\mathbf{X}^{\mathrm{T}}\mathbf{d} \tag{9.3.8}$$

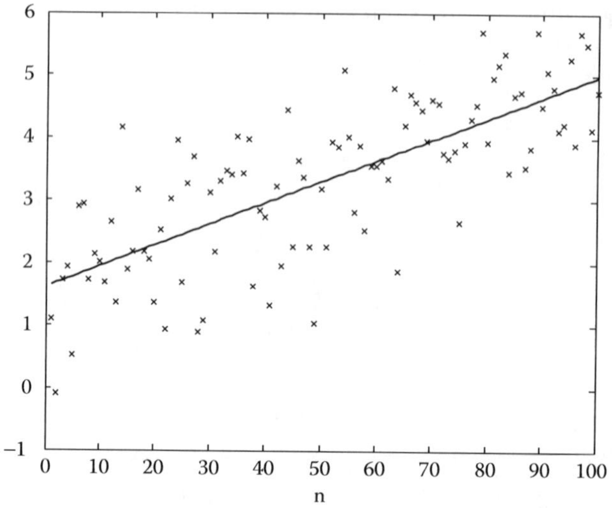

Figure 9.3.1 Illustration of Example 9.3.2.

and from the data shown in Figure 9.3.1

$$\hat{\mathbf{w}} = \begin{bmatrix} \hat{a} \\ \hat{b} \end{bmatrix} = \begin{bmatrix} 1.6147 \\ 0.0337 \end{bmatrix}$$

The straight line was also plotted to verify the procedure of LSE. The data were produced using the equation $d(n) = 1.5 + 0.035n + \text{randn}$ for $n = 1$ to 100.

9.4 *Orthogonality principle*

To obtain the orthogonality principle for the least-squared problem we follow the procedure developed for the Wiener filters. Therefore, using the unweighted sum of the squares of the error, we obtain

$$\frac{\partial J(w_1, w_2, \cdots, w_M)}{\partial w_k} = \frac{\partial}{\partial w_k}\left[\sum_{m=1}^{N} e(m)e(m)\right] = 2\sum_{m=1}^{N} e(m)\frac{\partial e(m)}{\partial w_k} \quad k = 1, 2, \cdots, M \tag{9.4.1}$$

But (9.2.6) is equal to ($\mathbf{w}$ has M coefficients)

$$e(m) = d(m) - \sum_{k=1}^{M} w_k x_k(m) \tag{9.4.2}$$

and, therefore, taking the derivative of $e(m)$ with respect to w_k and introducing the results in (9.4.1), we obtain

$$\frac{\partial J}{\partial w_k} = -2\sum_{m=1}^{N} e(m)x_k(m) \tag{9.4.3}$$

We note that when $\mathbf{w} = \hat{\mathbf{w}}$ (the optimum value) we have the relationship $\frac{\partial J}{\partial w_k} = 0$ for $k = 1, 2, \ldots, M$, and hence, (9.4.3) becomes

$$\sum_{m=1}^{N} \hat{e}(m)x_k(m) = \hat{e}\,\mathbf{x}_k \quad k = 1, 2, \cdots, M \tag{9.4.4}$$

where

$$\hat{\mathbf{e}} = [\hat{e}(1) \quad \hat{e}(2) \quad \hat{e}(3) \cdots \hat{e}(N)]^{\mathrm{T}} = \mathbf{d} - \hat{\mathbf{y}} \tag{9.4.5}$$

$$\mathbf{x} = [x_k(1) \quad x_k(2) \cdots x_k(N)]^{\mathrm{T}} \quad k = 1, 2, \cdots, M \tag{9.4.6}$$

the estimated error $\hat{e}(m)$ is optimum in the least-squares sense. The above result is known as the *principle of orthogonality.*

Corollary

Equations (9.2.6) may be written as the sum of the columns of **X** as follows

$$\hat{\mathbf{y}} = \sum_{k=1}^{M} \mathbf{x}_k(n)\hat{w}_k \qquad n = 1, 2, \cdots, N \tag{9.4.7}$$

Multiplying (9.4.7) by $\hat{\mathbf{e}}$ and taking into consideration, we obtain

$$\hat{\mathbf{e}}^{\mathrm{T}}\hat{\mathbf{y}} = 0 \tag{9.4.8}$$

The above corollary indicates that when the coefficients of the filter are optimum in the least-squares sense, then the output of the filter and the error are orthogonal.

Example 9.4.1: Using the results of Example 9.2.1, we find

$$\hat{\mathbf{e}} = \mathbf{d} - \hat{\mathbf{y}} = \begin{bmatrix} 1.05953819523825 \\ 0.91864528560697 \\ 0.48159639439564 \\ 1.21506254286554 \end{bmatrix}; \quad \hat{\mathbf{e}}^{\mathrm{T}}\mathbf{x}_0 = 2.505 \times 10^{-15}; \quad \hat{\mathbf{e}}^{\mathrm{T}}\mathbf{x}_1 = 1.457 \times 10^{-15}$$

9.5 Projection operator

Projection operator gives another from of interpretation to the solution of the least-squares problem. Let us, for clarity, assume that we have 2 (N vectors in the Nth dimensional case) vectors x_k that form two-dimensional subspace (see Figure 9.5.1). The vectors $\mathbf{x}_1$ and $\mathbf{x}_2$ constitute the column space of the data matrix $\mathbf{X}$.

We note the following:

1. The vector $\hat{\mathbf{d}}$ is obtained as a linear combination of the data column space $\mathbf{x}_1, \mathbf{x}_2, \cdots, \mathbf{x}_M$ of X that constitutes the subspace of M.
2. From all the vectors in the subspace spanned by $\mathbf{x}_1, \mathbf{x}_2, \cdots, \mathbf{x}_M$ the vector $\hat{\mathbf{d}}$ has the minimum Euclidian distance from $\mathbf{d}$.
3. The difference $\hat{\mathbf{e}} = \mathbf{d} - \hat{\mathbf{d}}$ is a vector that is orthogonal to the subspace.

We also note that $\hat{\mathbf{y}}$ satisfies the above three properties. From (9.4.7) we observe that $\hat{\mathbf{y}}$ is a linear combination of the data column space, which spans the subspace. Next, minimizing $\hat{\mathbf{e}}^T\hat{\mathbf{e}}$, where $\hat{\mathbf{e}} = \mathbf{d} - \hat{\mathbf{d}}$, is equivalent to minimizing the Euclidian distance between $\mathbf{d}$ and $\hat{\mathbf{y}}$. The third property is satisfied by (9.4.8). Therefore, we can conclude that $\hat{\mathbf{y}}$ is that the projection of $\mathbf{d}$ into the subspace spanned by the vectors $\mathbf{x}_1, \mathbf{x}_2, \cdots, \mathbf{x}_M$.

Equation (9.4.7) may also be written in the matrix form

$$\hat{\mathbf{y}} = \mathbf{X}\hat{\mathbf{w}} = \mathbf{X}(\mathbf{X}^T\mathbf{X})^{-1}\mathbf{X}^T\mathbf{d} \tag{9.5.1}$$

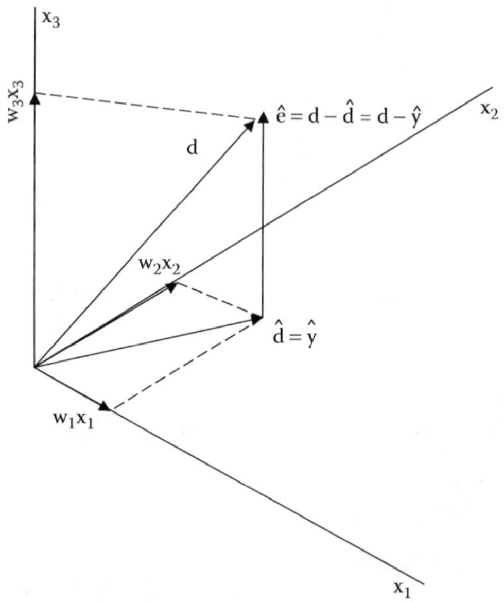

Figure 9.5.1 Vector space interpretation of the least-squares problems for $N = 3$ (data space) and $M = 2$ (estimation subspace).

where we set $\hat{\mathbf{w}} = \mathbf{R}^{-1}\mathbf{p}$ (see (9.2.21)), $\mathbf{R}^{-1} = (\mathbf{X}^T\mathbf{X})^{-1}$ (see (9.2.17)), and $\mathbf{p} = \mathbf{X}^T\mathbf{d}$ (see (9.2.18)). Since the matrix

$$\mathbf{P} = \mathbf{X}(\mathbf{X}^T\mathbf{X})^{-1}\mathbf{X}^T \tag{9.5.2}$$

projects the desired vector in the N-dimensional space to $\hat{\mathbf{y}}$ in the M-dimensional subspace ($N > M$), it is known as the *projection matrix* or *projection operator.* The name is due to the fact that the matrix **P** projects the data vector d onto the column space of **X** to provide the least-squares estimate $\hat{\mathbf{y}}$ of **d**.

The least-squares error can be expressed as

$$\hat{\mathbf{e}} = \mathbf{d} - \hat{\mathbf{y}} = \mathbf{d} - \mathbf{P}\mathbf{d} = (\mathbf{I} - \mathbf{P})\mathbf{d} \tag{9.5.3}$$

where **I** is an $N \times N$ identity matrix. The projection matrix is equal to its transpose (Hermitian for complex matrix) and *independent*, that is

$$\mathbf{P} = \mathbf{P}^T; \quad \mathbf{P}^2 = \mathbf{P}^T\mathbf{P} = \mathbf{P} \tag{9.5.4}$$

The matrix **I-P** is known as the *orthogonal complement projection operator.* The filter coefficients are given by

$$\hat{\mathbf{w}} = \mathbf{R}^{-1}\mathbf{p} = (\mathbf{X}^T\mathbf{X})^{-1}\mathbf{X}^T\mathbf{d} \tag{9.5.5}$$

where

$$\mathbf{X}^+ \equiv (\mathbf{X}^T\mathbf{X})^{-1}\mathbf{X}^T \tag{9.5.6}$$

is an $M \times N$ matrix known as the *pseudo-inverse* or the Moore-Penrose generalized inverse of matrix **X** (see Appendix A).

Example 9.5.1: Using the data given in Example 9.2.1, we obtain

$$\mathbf{P} = \mathbf{X}(\mathbf{X}^T\mathbf{X})^{-1}\mathbf{X}^T = \begin{bmatrix} 0.7278 & 0.2156 & 0.2434 & 0.3038 \\ -0.2156 & 0.7762 & 0.0013 & 0.3566 \\ 0.2434 & 0.0013 & 0.0890 & 0.1477 \\ 0.3038 & 0.3566 & 0.1477 & 0.4068 \end{bmatrix}$$

$$\hat{\mathbf{y}} = \mathbf{P}\mathbf{d} = [1.0595 \quad 0.9186 \quad 0.4815 \quad 1.2150]$$

$$\hat{\mathbf{e}} = (\mathbf{I} - \mathbf{P})\mathbf{d} = [0.0595 \quad 0.0813 \quad 0.5184 \quad -0.2150]^T$$

9.6 Least-squares finite impulse response filter

The error of the filter is given by

$$e(n) = d(n) - \sum_{k=1}^{M} w_k x(n-k) = d(n) - \mathbf{w}^{\mathrm{T}}\mathbf{x}(n) \tag{9.6.1}$$

where $d(n)$ is the desired signal,

$$\mathbf{x}(n) = [x(n) \quad x(n-1) \quad \cdots \quad x(n-M+1)]^{\mathrm{T}} \tag{9.6.2}$$

is the input data to the filter, and

$$\mathbf{w} = [w_1 \quad w_2 \cdots w_M]^{\mathrm{T}} \tag{9.6.3}$$

is the filter coefficient vector.

It turns out that the exact form of **e**, **d**, and **X** depends on the range $N_i \le n \le N_f$ of the data to be used. Therefore, the range of the square-error summation then becomes

$$J \triangleq E = \sum_{n=N_i}^{n=N_f} e^2(n) = \mathbf{e}^{\mathrm{T}}\mathbf{e} \tag{9.6.4}$$

The least-squares finite impulse response filter is found by solving the least-squares normal equations (see (9.2.20) and (9.2.30))

$$(\mathbf{X}^{\mathrm{T}}\mathbf{X})\,\hat{\mathbf{w}} = \mathbf{X}^{\mathrm{T}}\mathbf{d} = \mathbf{p} \quad (\text{or} \;\; \mathbf{R}\hat{\mathbf{w}} = \mathbf{p}) \tag{9.6.5}$$

with the minimum least-squares error

$$J_{\min} \triangleq E_{\min} = E_d - \mathbf{p}^{\mathrm{T}}\hat{\mathbf{w}} \tag{9.6.6}$$

where $E_d = \mathbf{d}^{\mathrm{T}}\mathbf{d}$ is the energy of the desired signal. The elements of the time-averaged correlation matrix **R** are given by (the real averaged correlation coefficients must be divided by $N_f - N_i$)

$$r_{ij} = \mathbf{x}_i^{\mathrm{T}}\mathbf{x}_j = \sum_{n=N_i}^{N_f} x(n+1-i)x(n+1-j) \quad 1 \le i,\, j \le M \tag{9.6.7}$$

There are two important different ways to select the summation range $N_i \le n \le N_f$, which are exploited in Problem 9.6.1. These are the *no-window* case, where $N_i = M - 1$ and $N_f = N - 1$, and the full-window case, where the range of the summation is from $N_i = 0$ to $N_f = N + M - 2$. The no-window method is also known as the *autocorrelation* method, and the full-window method is also known as the *covariance method.*

The covariance method *data matrix* **D** is written as follows:

$$\mathbf{D}^{\mathrm{T}} = [\mathbf{x}(M)\ \ \mathbf{x}(M+1)\ \cdots\ \mathbf{x}(M)] = \begin{bmatrix} x(M) & x(M+1) & \cdots & x(N) \\ x(M-1) & x(M) & \cdots & x(N-1) \\ \vdots & \vdots & & \vdots \\ x(1) & x(2) & \cdots & x(N-M+1) \end{bmatrix} \tag{9.6.8}$$

Then the $M \times M$ time-averaged correlation matrix is given by

$$\mathbf{R} = \sum_{n=M}^{N} \mathbf{x}(n)\mathbf{x}^{\mathrm{T}}(n) = \mathbf{D}^{\mathrm{T}}\mathbf{D} \tag{9.6.9}$$

Book MATLAB function for covariance data matrix

```
function[dT]=aadatamatrixcovmeth(x,M)
%function[dT]=aadatamatrixconvmeth(x,M)
%M=number of filter coefficients;x=data vector;
%dT=transposed data matrix;
for m=1:M
  for n=1:length(x)-M+1
    dT(m,n)=x(M-m+n);
  end;
end;
```

Example 9.6.1: If the data vector is x = [0.7 1.4 0.4 1.3 0.1]^{T} and the data filter has three coefficients, then

$$\mathbf{D}^{\mathrm{T}} = \begin{bmatrix} 0.4 & 1.3 & 0.1 \\ 1.4 & 0.4 & 1.3 \\ 0.7 & 1.4 & 0.4 \end{bmatrix} \qquad (\mathbf{D}^{\mathrm{T}} \text{ is Toeplitz matrix})$$

$$\mathbf{R} = \mathbf{D}^{\mathrm{T}}\mathbf{D} = \begin{bmatrix} 1.8600 & 1.2100 & 2.1400 \\ 1.2100 & 3.8100 & 2.0600 \\ 2.1400 & 2.0600 & 2.6100 \end{bmatrix}$$

The data matrix in (9.6.8) has the following properties:

Property 1: The correlation matrix **R** is equal to its transpose (for complex quantities, **R** is Hermitian $\mathbf{R} = \mathbf{R}^T$). The proof is directly found from (9.6.9).

Property 2: The correlation matrix is nonnegative definite, $\mathbf{a}^T\mathbf{R}\mathbf{a} \geq 0$ for any $M \times 1$ vector a (see Problem 9.6.1).

Property 3: The eigenvalues of the correlation matrix **R** are all real and nonnegative (see Section 5.1).

Property 4: The correlation matrix is the product of two rectangular Toeplitz matrices that are the transpose of each other (see Example 9.6.1)

The following book MATLAB function will produce the results for the no-window method FIR filter:

Book MATLAB function no-window LS method

```
function[R,w,Jmin]=aanowindowleastsqufir(x,M,d)
%x=data of length N;M=number of filter coefficient;
%d=desired signal=[d(M)    d(M+1)    ...    d(N)]';
N=length(x);
for i=1:M
  for j=1:N-M+1
      D(i,j)=x(M-i+j);
  end;
end;
Dt=D';
R=D*Dt;
p=D*d(1,1:N-M+1)';
w=inv(R)*p;
Jmin=d'*d-p'*w;
```

9.7 Introduction to RLS algorithm

The least-squares solution (9.2.21) is not very practical in the actual implementation of adaptive filters. This is true, because we must know all the past samples of the input signal, as well as the desired output must be available at every iteration. The RLS algorithm is based on the LS estimate of the filter coefficients $\mathbf{w}(n-1)$ at iteration $n-1$, by computing its estimate at iteration n using the newly arrived data. This type of algorithm is known as the *recursive least-squares (RLS) algorithm*. This algorithm may be viewed as a special case of the Kalman filter.

To implement the recursive method of least squares, we start the computation with known initial conditions and then update the old estimate based on the information contained in the new data samples. Next, we minimize the cost function $J(n)$, where n is the variable length of the observed data. Hence, we write (9.2.3) in the form

$$J(n) = \sum_{k=1}^{n} \eta_n(k) e^2(k), \quad \eta_n(k) \equiv \text{weighting factor} \tag{9.7.1}$$

where

$$e(k) = d(k) - y(k) = d(k) - \mathbf{w}^{\mathrm{T}}(k)\mathbf{x}(k) \tag{9.7.2}$$

$$\mathbf{x}(k) = [x(k) \quad x(k-1) \quad \cdots \quad x(k-M+1)]^{\mathrm{T}} \tag{9.7.3}$$

$$\mathbf{w}(n) = [w_1(n) \quad w_2(n) \quad \cdots \quad w_M(n)]^{\mathrm{T}} \tag{9.7.4}$$

Note that the filter coefficient are fixed during the observation time $1 \le k \le n$ during which the cost function $J(n)$ is defined.

In standart RLS algorithm, the weighting factor $\eta_n(k)$ is chosen to have the exponential form

$$\eta_n(k) = \lambda^{n-k} \quad k = 1, 2, \cdots, n \tag{9.7.5}$$

where the value of λ is less than one and, hence, $\eta_n(k)$ is confined in the range $0 < \eta_n(k) \le 1$ for k = 1, 2,..., n. The weighting factor λ is also known as the *forgetting factor,* since it weights (emphasizes) the recent data and tends to forget the past. This property helps in producing an adaptive algorithm with some tracking capabilities. Therefore, we must minimize the cost function

$$J(n) = \sum_{k=1}^{n} \lambda^{n-k} e^2(k) \tag{9.7.6}$$

The minimum value of $J(n)$ is attained (see Section 9.2) when the *normal equations* (see (9.2.20)

$$\mathbf{R}_\lambda(n)\hat{\mathbf{w}} = \mathbf{p}_\lambda(n) \quad (\hat{\mathbf{w}} = \mathbf{R}_\lambda^{-1}\mathbf{p}_\lambda(n)) \tag{9.7.7}$$

are satisfied and where the $M \times M$ correlation matrix $\mathbf{R}_\lambda(n)$ is defined by (see Problem 9.7.1)

$$\mathbf{R}_\lambda(n) = \sum_{k=1}^{n} \lambda^{n-k} \mathbf{x}(k)\mathbf{x}^{\mathrm{T}}(k) = \mathbf{X}^{\mathrm{T}}\Lambda\mathbf{X} \tag{9.7.8}$$

and

$$\mathbf{p}_\lambda(n) = \sum_{k=1}^{n} \lambda^{n-k}\mathbf{x}(k)d(k) = \mathbf{X}^T\Lambda\mathbf{d}; \quad \Lambda = \text{diag}[\lambda^{n-1} \quad \lambda^{n-2} \cdots 1] \tag{9.7.9}$$

Note that $\mathbf{R}_\lambda(n)$ differs from $\mathbf{R}$ in the following two respects: 1) the common matrix $\mathbf{x}(k)\mathbf{x}^T(k)$ is weighted by the exponential factor λ^{n-k}, 2) the use of prewindowing is assumed, according to which the input data prior to time $k = 1$ are zero and, thus, $k = 1$ becomes the lower limit of the summation; the same is true for $\mathbf{p}_\lambda(n)$.

The minimum total square error is (see Problem 9.3.2)

$$J_{\min} = \mathbf{d}^T(n)\Lambda\mathbf{d}(n) - \mathbf{w}^T(n)\mathbf{p}_\lambda(n) = \sum_{k=1}^{n} \lambda^{n-k}d^2(k) - \mathbf{w}^T(n)\mathbf{p}_\lambda(n) \tag{9.7.10}$$

Next, we wait for a time such that $n > M$, where in practice $\mathbf{R}_\lambda$ is nonsingular, and then compute $\mathbf{R}_\lambda$ and $\mathbf{p}_\lambda(n)$. Next we solve the normal Equations (9.7.7) to obtain the filter coefficients $\hat{\mathbf{w}}(n)$. This is repeated with the arrival of the new pairs $\{\mathbf{x}(\mathrm{n}), d(n)\}$, that is, at times $n + 1$, $n + 2$, ….

If we isolate the term at $k = n$, we can write (9.7.7) in the form

$$\mathbf{R}_\lambda(n) = \lambda\left[\sum_{k=1}^{n-1} \lambda^{n-1-k}\mathbf{x}(k)\mathbf{x}^T(k)\right] + \mathbf{x}(n)\mathbf{x}^T(n) \tag{9.7.11}$$

By definition the expression in the bracket is $\mathbf{R}_\lambda(n-1)$, and thus, (9.7.11) becomes

$$\mathbf{R}_\lambda(n) = \lambda\mathbf{R}_\lambda(n-1) + \mathbf{x}(n)\mathbf{x}^T(n) \tag{9.7.12}$$

The above equation shows that the "new" correlation matrix $\mathbf{R}_\lambda(n)$ is updated by weighting the "old" correlation matrix $\mathbf{R}_\lambda(n-1)$ with the factor and adding the correlation term $\mathbf{x}(n)\mathbf{x}^T(n)$.

Similarly, using (9.7.9) we obtain

$$\mathbf{p}_\lambda(n) = \lambda\mathbf{p}_\lambda(n-1) + \mathbf{x}(n)d(n) \tag{9.7.13}$$

which gives an update of the cross-correlation vector.

Next, we try to find $\hat{\mathbf{w}}$ by iteration and thus avoid solving the normal Equations (9.7.7).

The matrix inversion lemma

There are several relations, which are known as the inversion lemma. Let **A** be an $M \times M$ invertible matrix and **x** and **y** be two $M \times 1$ vectors such that $(\mathbf{A} + \mathbf{x}\mathbf{y}^T)$ is invertible. Then we have (see Problem 9.7.3)

$$(\mathbf{A} + \mathbf{x}\mathbf{y}^T)^T = \mathbf{A}^{-1} - \frac{\mathbf{A}^{-1}\mathbf{x}\mathbf{y}^T\mathbf{A}^{-1}}{1 + \mathbf{y}^T\mathbf{A}^{-1}\mathbf{x}} \tag{9.7.14}$$

Next, let **A** and **B** be positive-definite $M \times M$ matrices related by

$$\mathbf{A} = \mathbf{B}^{-1} + \mathbf{C}\mathbf{D}^{-1}\mathbf{C}^T \tag{9.7.15}$$

where **D** is a positive-definite matrix of $N \times M$ and **C** is another $M \times N$ matrix. The inversion lemma tell us that (see Problem 9.7.4)

$$\mathbf{A}^{-1} = \mathbf{B} - \mathbf{B}\mathbf{C}(\mathbf{D} + \mathbf{C}^T\mathbf{B}\mathbf{C})^{-1}\mathbf{C}^T\mathbf{B} \tag{9.7.16}$$

Furthermore, (9.7.14) can also be in the form

$$(\mathbf{A} + a\mathbf{x}\mathbf{x}^T)^{-1} = \mathbf{A}^{-1} - \frac{a\mathbf{A}^{-1}\mathbf{x}\mathbf{x}^T\mathbf{A}^{-1}}{1 + \mathbf{x}^T\mathbf{A}^{-1}\mathbf{x}} \tag{9.7.17}$$

$$(\lambda\mathbf{A} + \mathbf{x}\mathbf{x}^T)^{-1} = \lambda^{-1}\mathbf{A}^{-1} - \frac{(\lambda^{-1}\mathbf{A}^{-1})\mathbf{x}\mathbf{x}^T(\lambda^{-1}\mathbf{A}^{-1})}{1 + \lambda^{-1}\mathbf{x}^T\mathbf{A}^{-1}\mathbf{x}} \tag{9.7.18}$$

The RLS algorithm

To evaluate the inverse of $\mathbf{R}_\lambda(n)$ we set $\mathbf{A} = \lambda\mathbf{R}_\lambda(n-1)$ and comparing (9.7.12) and (9.7.18) we find

$$\mathbf{R}_\lambda^{-1}(n) = \lambda^{-1}\mathbf{R}_\lambda^{-1}(n-1) - \frac{\lambda^{-2}\mathbf{R}_\lambda^{-1}(n-1)\mathbf{x}(n)\mathbf{x}^T(n)\mathbf{R}_\lambda^{-1}(n-1)}{1 + \lambda^{-1}\mathbf{x}^T(n)\mathbf{R}_\lambda^{-1}(n-1)\mathbf{x}(n)} \tag{9.7.19}$$

The same relation is found if we set $\mathbf{A} = \mathbf{R}_\lambda(n)$, $\mathbf{B}^{-1} = \lambda\mathbf{R}_\lambda(n-1)$, $\mathbf{C} = \mathrm{x}(n)$, and $\mathbf{D} = 1$ in (9.7.16). Next we define the column vector $\mathbf{g}(n)$ as follows:

$$\mathbf{g}(n) = \frac{\lambda^{-1}\mathbf{R}_\lambda^{-1}(n-1)\mathbf{x}(n)}{1 + \lambda^{-1}\mathbf{x}^T(n)\mathbf{R}_\lambda^{-1}(n-1)\mathbf{x}(n)} = \frac{\mathbf{R}_\lambda^{-1}(n-1)\mathbf{x}(n)}{\lambda + \mathbf{x}^T(n)\mathbf{R}_\lambda^{-1}(n-1)\mathbf{x}(n)} \tag{9.7.20}$$

This vector is known as the *gain vector.*

Comparing (9.7.20) and (9.7.19) we obtain

$$\mathbf{R}_\lambda^{-1}(n) = \lambda^{-1}[\mathbf{R}_\lambda^{-1}(n-1) - \mathbf{g}(n)\mathbf{x}^T(n)\mathbf{R}_\lambda^{-1}(n-1)] \tag{9.7.21}$$

which is known as the conventional RLS (CRLS) algorithm. This algorithm is valid for both linear combiner and FIR filters because no assumption is made about the input data vector $\mathbf{x}(n)$. For FIR filter we generally assume prewindowing, $x(-1) = 0$ or equivalently $x(n) = 0$ for $-M \le n \le -1$.

By rearranging (9.7.21) we obtain

$$\mathbf{g}(n) = \lambda^{-1}[\mathbf{R}_\lambda^{-1}(n-1) - \mathbf{g}(n)\mathbf{x}^T(n)\mathbf{R}_\lambda^{-1}]\mathbf{x}(n) \tag{9.7.22}$$

Using (9.7.21) and (9.7.22) we find the relation

$$\mathbf{g}(n) = \mathbf{R}_\lambda^{-1}(n)\mathbf{x}(n) \tag{9.7.23}$$

Next, we substitute (9.7.13) in (9.7.7), and considering (9.7.23), we find

$$\hat{\mathbf{w}} = \lambda\mathbf{R}_\lambda^{-1}(n)\mathbf{p}_\lambda(n) + \mathbf{R}_\lambda^{-1}(n)\mathbf{x}(n)d(n) = \lambda\mathbf{R}_\lambda^{-1}(n)\mathbf{p}_\lambda(n)\mathbf{g}(n)d(n) \tag{9.7.24}$$

Our next step is to introduce the value of $\mathbf{R}_\lambda^{-1}(n)$ from (9.7.21) in (9.7.24) to find the relation

$$\hat{\mathbf{w}} = \hat{\mathbf{w}}(n-1) + \mathbf{g}(n)[d(n) - \hat{\mathbf{w}}^T(n-1)\mathbf{x}(n)] \tag{9.7.25}$$

where we used the following two relations: $\hat{\mathbf{w}}(n-1) = \mathbf{R}_\lambda^{-1}(n-1)\mathbf{p}(n-1)$, $\mathbf{x}^T(n)\hat{\mathbf{w}}(n-1) = \hat{\mathbf{w}}^T(n-1)\mathbf{x}(n)$. If we define the *a priori estimation error* by

$$\hat{\mathbf{e}}_{n-1}(n) = d(n) - \hat{\mathbf{w}}^T(n-1)\mathbf{x}(n) \tag{9.7.26}$$

(9.7.25) takes the final form

$$\hat{\mathbf{w}}(n) = \hat{\mathbf{w}}(n-1) + \mathbf{g}(n)\,\hat{\mathbf{e}}_{n-1}(n) \tag{9.7.27}$$

The inner product $\hat{\mathbf{w}}^T(n-1)\mathbf{x}(n)$ represents an estimate of the desired response $d(n)$, based on the previous least-squares estimate of the vector coefficients. The last two equations suggest a block-diagram representation shown in Figure 9.7.1.

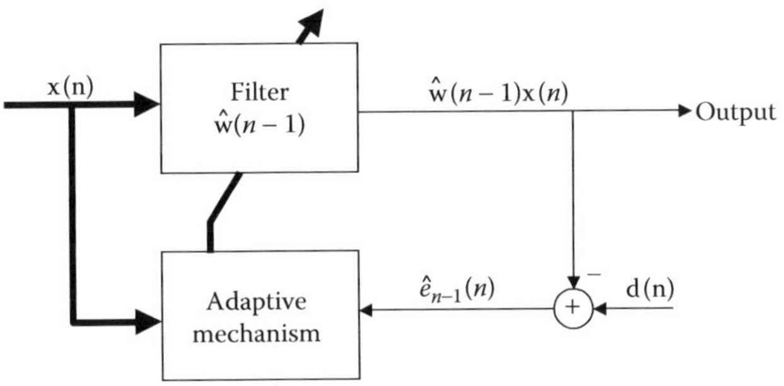

Figure 9.7.1 RLS block-diagram algorithm.

The *posteriori estimation error* is given by

$$\hat{e}_n(n) = d(n) - \hat{\mathbf{w}}^T(n)\mathbf{x}(n) \tag{9.7.28}$$

where the current least-squares estimates of the filter coefficients are used.

Initialization of the RLS algorithm

To start the algorithm we must introduce proper initializations of $\mathbf{R}_\lambda(0)$ and $\hat{\mathbf{w}}(0)$. For values of $n < M$, the matrix (see (9.7.8))

$$\mathbf{R}_\lambda(n) = \sum_{k=1}^{n} \lambda^{n-k}\mathbf{x}(k)\mathbf{x}^T(k) \tag{9.7.29}$$

has rank that is less than its dimension M. Hence, the inverse $\mathbf{R}_\lambda(n)$ does not exist for $n < M$. To remedy the situation, we modify (9.7.29) as follows

$$\mathbf{R}_\lambda(n) = \sum_{k=1}^{n} \lambda^{n-k}\mathbf{x}(k)\mathbf{x}^T(k) + \delta\lambda^n\mathbf{I} \tag{9.7.30}$$

where

$$\mathbf{R}_\lambda(0) = \delta\mathbf{I} \tag{9.7.31}$$

and $\mathbf{I}$ is an $M \times M$ identity matrix, δ is a small number, and $\lambda < 1$. As n increases, the effect of the initial condition diminishes and has negligible effect on the steady-state value. It has been suggested that δ must be less than $0.01\sigma_x^2$, where σ_x^2 is the variance of the data.

It is common in practice to set the initial value of the filter coefficients equal to zero

$$\hat{\mathbf{w}}(0) = \mathbf{0} \tag{9.7.32}$$

where $\mathbf{0}$ is an $M \times 1$ zero vector.

Summary of the RLS algorithm

To simplify computations, we define the following intermediate vector

$$\bar{\mathbf{g}}(n) = \mathbf{R}_\lambda^{-1}(n-1)\mathbf{x}(n) \tag{9.7.33}$$

Then the gain vector, after being multiplied by λ, becomes (see (9.7.1))

$$\mathbf{g}(n) = \frac{1}{\lambda + \mathbf{x}^T(n)\bar{\mathbf{g}}(n)}\bar{\mathbf{g}}(n) \tag{9.7.34}$$

Therefore (9.7.19) becomes

$$\mathbf{R}_\lambda^{-1}(n) = \lambda^{-1}\left[\mathbf{R}_\lambda^{-1}(n-1) - \mathbf{g}(n)\bar{\mathbf{g}}^T(n)\right] \tag{9.7.35}$$

The above equation states that, given the $(n-1)^{th}$ matrix $\mathbf{R}_\lambda^{-1}(n-1)$ and the new observation $\{\mathbf{x}(n), d(n)\}$, we can compute the new matrix $\mathbf{R}_\lambda^{-1}(n)$ following the procedure that is given in Table 9.7.1.

In practice, a potential problem of the RLS algorithm is the accumulation of the round-off error. Engineers who use adaptive filtering algorithms should be aware of this potential problem and, therefore, they must test the algorithm before final implementation.

Example 9.7.1: Figure 9.7.2 shows a system identification setup with a coloring input filter, which has an impulse response $h_1(n)$. This filter colors (correlates) its input $g(n)$ and produces white (uncorrelated) signal at its output $x(n)$. Figure 9.7.3 shows the learning curve of the RLS algorithm for this system modeling problem using the following values of parameters:

No. of iterations = 250, variance of the plant noise $\sigma_v^2 = 0.001$, plant impulse response h = [0.1 0.2 0.3 0.4 0.5 0.4 0.3 0.2 0.1], length of the adaptive filter N = 10, coloring filter impulse response h_1 = [0.5 1 −0.5], variance of the coloring filter input $\sigma_g^2 = 1$, forgetting factor $\lambda = 0.97$, the parameter $\delta = 0.001$ and the number of runs to be used for ensemble averaging = 50.

Table 9.7.1 Summary of the Conventional Recursive LS (CRLS) Algorithm

Initialization: $\mathbf{R}_\lambda^{-1}(0) = \delta^{-1}\mathbf{I}, \quad \delta < 0.01\sigma_x^2, \quad \sigma_x^2 = \text{variance of the data}$

$\hat{\mathbf{w}}(0) = \mathbf{0}$

Input: Filter coefficient vector estimate $\hat{\mathbf{w}}(n-1)$
Input vector $\mathbf{x}(n)$
Design output $\mathbf{d}(n)$
Matrix $\mathbf{R}_\lambda^{-1}(n-1)$

Output Filter output $\hat{y}_{n-1}(n)$

Update filter coefficient vector $\hat{\mathbf{w}}(n)$

Update matrix $\mathbf{R}_\lambda^{-1}(n)$

Algorithm:

1. Gain vector computation $\bar{\mathbf{g}}(n) = \mathbf{R}_\lambda^{-1}(n-1)\mathbf{x}(n), \mathbf{g}(n) = \dfrac{1}{\lambda + \mathbf{x}^{\mathrm{T}}(n)\bar{\mathbf{g}}(n)}\bar{\mathbf{g}}(n)$

2. Output (filtering) $\hat{y}_{n-1}(n) = \hat{\mathbf{w}}^{\mathrm{T}}(n-1)\mathbf{x}(n)$

3. Error estimate $\hat{e}_n(n) = d(n) - \bar{y}_{n-1}(n) = d(n) - \hat{\mathbf{w}}^{\mathrm{T}}(n)\mathbf{x}(n)$

4. Coefficient vector updating $\hat{\mathbf{w}}(n) = \hat{\mathbf{w}}(n-1) + \mathbf{g}(n)\hat{e}_{n-1}(n)$

5. $\mathbf{R}_\lambda^{-1}(n)$ update

$$\mathbf{R}_\lambda^{-1}(n) = \lambda^{-1}[\mathbf{R}_\lambda^{-1}(n-1) - \mathbf{g}(n)[\mathbf{x}^{\mathrm{T}}(n)\mathbf{R}_\lambda^{-1}(n-1)]] = \lambda^{-1}[\mathbf{R}_\lambda^{-1}(n-1) - \mathbf{g}(n)\bar{\mathbf{g}}^{\mathrm{T}}(n)]$$

Note: To update $\mathbf{R}_\lambda^{-1}$ in practical applications we only compute its upper or low triangular part and then determine the other part by adding its transpose (Hermitian in complex case).

The Book MATLAB RLS program is given below:

```
function [w,J]=aarls(L,var2,h,N,h1,lambda,delta,I);
%L=No. of iterations;
%var2=Variance of the plant noise;
%h=Plant impulse response;
%N=Length of the Adaptive Filter;
%h1=Coloring filter impulse response;
%lambda=Forgetting factor in the RLS algorithm;
%delta=Parameter in the RLS algorithm;
%I=No. of runs to be used for ensemble averaging;
J=zeros(L,1);
for k=1:I
  x1=randn(L,1);x=filter(h1,1,x1);
  v=sqrt(var2)*randn(L,1); d=filter(h,1,x)+v;
       w=zeros(N,1);     xd=zeros(N,1);
       R_inv=(1/delta)*eye(N);
```

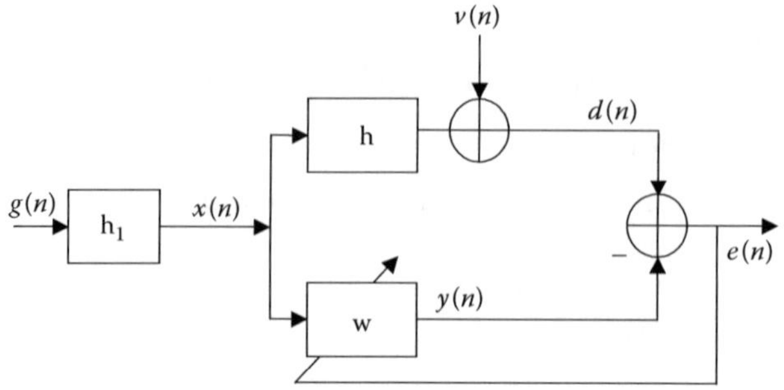

Figure 9.7.2 Adaptive system modeling with correlated input.

```
        for n=1:L
                xd=[x(n);xd(1:length(xd)-1)];
                gbar=R_inv*xd;
                g=gbar/(lambda+xd'*gbar);
                y=w'*xd;
                e=d(n)-y;
                w=w+g*e;
                R_inv=(1/lambda)*(R_inv-g*gbar');
                J(n)=J(n)+e^2;
          end;
end;
J=J/I;
nn=0:L-1;plot(nn,10*log10(J));
xlabel('Iteration no.');ylabel('MSE in dB');
```

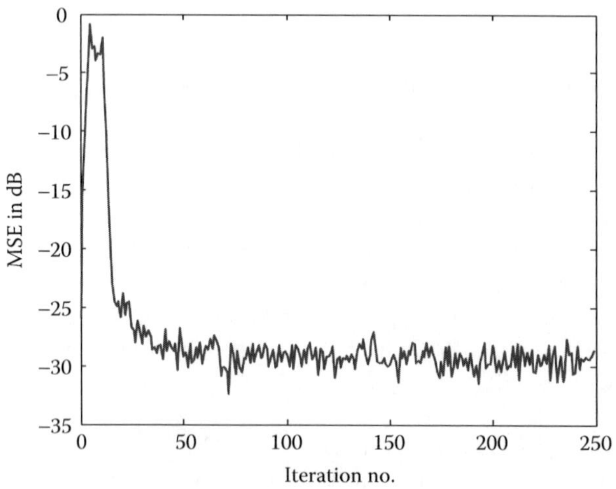

Figure 9.7.3 Learning curve of the RLS algorithm in Example 9.7.1.

Problems

9.2.1 Verify (9.2.20).

9.2.2 Find the vector coefficient of the filter and the minimum error if the following are given:

$$\mathbf{d} = [1\ 1\ 1\ 1]^{T};\ \mathbf{x}_1 = [0.7\ 1.4\ 0.4\ 1.3]^{T};\ \mathbf{x}_2 = [1.2\ 0.6\ 0.5\ 1.1]^{T};$$

$$\mathbf{x}_3 = [0.9\ 1.1\ 0.95\ 1.05]^{T}.$$

9.2.3 The time-averaged correlation matrix $\mathbf{R} = \mathbf{X}^T\mathbf{X}$ (must be divided by N to be truly time averaged) is invertible if, and only if, the columns x_k of X are linearly independent, or, equivalently, if and only if $\mathbf{R}$ is positive definite.

9.3.1 Let us assume that the signal from a LTI system is a constant, $y(n) = a$, and we observe the data $x(n)$, for $n = 1, 2, \ldots, N$. Find the estimate value of a, $\hat{a}$.

9.3.2 Verify (9.3.4).

9.6.1 Find the time-average autocorrelation matrices for the two different summation cases (no window and full window) for the range, $N_i \le n \le N_f$.

9.7.1 Verify (9.7.8).

9.7.2 Verify (9.7.10).

9.7.3 Verify (9.7.14).

9.7.4 Verify (9.7.15).

9.7.5 Find the signal-flow diagram for Equation (9.7.26) and Equation (9.7.27).

Hints-solutions-suggestions

9.2.1:
For simplicity and without loss of generality, we assume a 2×2 matrix and 2×1 vector.

$$\frac{\partial J}{\partial \mathbf{w}} = \frac{\partial}{\partial \mathbf{w}}[E_y - 2\mathbf{p}\mathbf{w}^T + \mathbf{w}^T\mathbf{R}\mathbf{w}] = \frac{\partial E_y}{\partial \mathbf{w}} - 2\mathbf{p}^T\frac{\partial \mathbf{w}}{\partial \mathbf{w}} + \frac{\partial}{\partial \mathbf{w}}(\mathbf{w}^T\mathbf{R}\mathbf{w}) = \mathbf{0};$$

$$-2\begin{bmatrix} \dfrac{\partial}{\partial w_1}(p_1 w_1 + p_2 w_2) \\ \dfrac{\partial}{\partial w_2}(p_1 w_1 + p_2 w_2) \end{bmatrix} + \begin{bmatrix} \dfrac{\partial}{\partial w_1}(r_{11}w_1^2 + r_{12}w_2 w_1 + r_{21}w_2 w_1 + r_{22}w_2^2) \\ \dfrac{\partial}{\partial w_2}(r_{11}w_1^2 + r_{12}w_2 w_1 + r_{21}w_2 w_1 + r_{22}w_2^2 \end{bmatrix}$$

$$= -2\begin{bmatrix} p_1 \\ p_2 \end{bmatrix} + \begin{bmatrix} 2r_{11}w_1 + 2r_{12}w_2 \\ 2r_{21}w_1 + 2r_{22}w_2 \end{bmatrix} = \mathbf{0}; \quad or \quad \mathbf{R}\hat{\mathbf{w}} = \mathbf{p}$$

9.2.2:
Hint: Follow Example 9.2.1.

9.2.3:
If the columns of **X** are linearly independent, then for every $\mathbf{y} \neq \mathbf{0}$, we have $\mathbf{X}\mathbf{y} \neq \mathbf{0}$. Hence, $\mathbf{y}^T(\mathbf{X}^T\mathbf{X})\mathbf{y} = (\mathbf{X}\mathbf{y})^T\mathbf{X}\mathbf{y} = \|\mathbf{X}\mathbf{y}\|^2 > 0$, which indicates that $\mathbf{R} = \mathbf{X}^T\mathbf{X}$ is positive definite and, therefore, nonsingular. If the columns of X are linearly dependent, then there is a $\mathbf{y}_0 \neq \mathbf{0}$. Therefore, $\mathbf{X}^T\mathbf{X}\mathbf{y}_0 = 0$ and **R** is singular.

9.3.1:

$$J(a) = \sum_{n=1}^{N} (x(n) - a)^2 \Rightarrow \frac{\partial J(a)}{\partial a} = \sum_{n=1}^{N} (-)2(x(n) - \hat{a}) = 0 \quad or \quad \sum_{n=1}^{N} x(n) - N\hat{a} = 0$$

$$or \quad \hat{a} = \frac{1}{N}\sum_{n=1}^{N} x(n), \text{ which is the sample mean estimator.}$$

9.3.2:

$$J_{\min} = J(\hat{w}) = \sum_{n=1}^{N} (d(n) - \hat{w}\, x(n))(d(n) - \hat{w}\, x(n))$$

$$= \sum_{n=1}^{N} d(n)(d(n) - \hat{w}\, x(n)) - \hat{w}\sum_{n=1}^{N} x(n)(d(n) - \hat{w}x(n))$$

$$= \sum_{n=1}^{N} d^2(n) - \hat{w}\sum_{n=1}^{N} d(n)x(n) - \hat{w}\sum_{n=1}^{N} d(n)x(n) + \hat{w}^2\sum_{n=1}^{N} x^2(n)$$

$$= \sum_{n=1}^{N} d^2(n) - \hat{w}\left(\sum_{n=1}^{N} d(n)x(n) + \sum_{n=1}^{N} d(n)\, x(n) - \frac{\sum_{n=1}^{N} d(n)x(n)}{\sum_{n=1}^{N} x^2(n)}\sum_{n=1}^{N} x^2(n)\right)$$

$$= \sum_{n=1}^{N} d^2(n) - \hat{w}\sum_{n=1}^{N} d(n)x(n).$$

9.6.1:
Case I (no window): Let us define the limits of interest as $N_i = M - 1$ and $N_f = N - 1$. In this case, no assumption is made about the data outside the interval $[1, N]$. Let us further assume that we measured the data (see Example 9.2.1)

$\mathbf{x} = [0.7\ \ 1.4\ \ 0.4\ \ 1.3\ \ 0.1]^T$, and the filter has three coefficients $\mathbf{w} = [w_1\ w_2\ w_3]$. For this case $N_i = M - 1 = 2$ and $N_f = N - 1 = 4$, and therefore (see (9.6.7))

$$r(i,j) = \mathbf{x}_i^T \mathbf{x}_j = \sum_{n=2}^{4} x(n+1-i)x(n+1-j) \quad 1 \le i, j \le 3$$

or

$$\mathbf{R} = \begin{bmatrix} 1.8600 & 1.2100 & 2.1400 \\ 1.2100 & 3.8100 & 2.0600 \\ 2.1400 & 2.0600 & 2.6100 \end{bmatrix} = \mathbf{R}^T.$$

The no-window method is known as the *autocorrelation* method. Case II (full window): In this method the range of summation is for $N_i = 0$ to $N_f = N + M - 2$. This is equivalent to setting equal to zero the samples $x(n - M)$, $x(2 - M)$, …, $x(1)$ and the samples $x(N)$, …, $x(N + M - 2)$. Hence, to calculate with MATLAB we write:

$\mathbf{x} = [0\ \ 0\ \ 0\ \ 0.7\ \ 1.4\ \ 0.4\ \ 1.3\ \ 0.1\ \ 0\ \ 0]^T = [\text{zeros}(1, M)\ \ x\ \ \text{zeros}(N, N + M - 2)]^T$

and, hence,

$$\mathbf{R} = \begin{bmatrix} 4.3100 & 2.1900 & 2.1400 \\ 2.1900 & 4.3100 & 2.1900 \\ 2.1400 & 2.1900 & 4.3100 \end{bmatrix}$$

Note that the matrix is *Toeplitz*.

9.7.1:
$y_n(k) = \mathbf{w}^T(n)\mathbf{x}(k)$ for $k = 1, 2, \ldots, n$ (1); $e_n(k) = d(k) - y_n(k)$ (2); $\mathbf{d}(n) = [d(1)\ d(2)\ d(n)]^T$ (3); $\mathbf{y}(n) = [y_n(1)\ \ y_n(2)\ \ \cdots\ \ y_n(n)]^T$ (4); $\mathbf{e}(n) = [e_n(1)\ \ e_n(2)\ \ \cdots\ \ e_n(n)]^T$ (5). Define $X^T = [\mathbf{x}(1)\ \ \mathbf{x}(2)\ \ \ldots\ \ \mathbf{x}(n)]$ (6). Then using (1) through (6) we obtain

$$e_n(1) = d(1) - y_n(1) = d(1) - \mathbf{w}^T(n)\mathbf{x}(1)$$

$$e_n(2) = d(2) - y_n(2) = d(2) - \mathbf{w}^T(n)\mathbf{x}(2)$$

$$\vdots$$

$$e_n(n) = d(n) - y_n(n) = d(n) - \mathbf{w}^T(n)\mathbf{x}(n)$$

or

$$e(n) = \mathbf{d}(n) - \mathbf{w}^{\mathrm{T}}(n)\mathbf{X}^{\mathrm{T}} = \mathbf{d}(n) - \mathbf{X}\mathbf{w}(n) \quad (7)$$

$$y(n) = \mathbf{X}\mathbf{w}(n) = \mathbf{w}^{\mathrm{T}}(n)\mathbf{X}^{\mathrm{T}} \quad (8)$$

$$J(n) = \mathbf{e}^{\mathrm{T}}(n)\Lambda\mathbf{e}(n) \qquad (9);$$

$$\Lambda = \begin{bmatrix} \lambda^{n-1} & 0 & \cdots & 0 \\ 0 & \lambda^{n-2} & \cdots & 0 \\ \vdots & & & \vdots \\ 0 & 0 & \cdots & 1 \end{bmatrix} \qquad (10)$$

From (7), (8), and (9) we obtain

$$J(n) = [\mathbf{d}(n) - \mathbf{X}\mathbf{w}(n)]^{\mathrm{T}}\Lambda[\mathbf{d}(n) - \mathbf{X}\mathbf{w}(n)]$$

$$= \mathbf{d}^{\mathrm{T}}(n)\Lambda\mathbf{d}(n) - 2\mathbf{d}^{\mathrm{T}}(n)\Lambda\mathbf{X}\mathbf{w}(n) + \mathbf{w}^{\mathrm{T}}(n)\mathbf{X}^{\mathrm{T}}\Lambda\mathbf{X}\mathbf{w}(n) \quad (11).$$

Following Problem 9.2.1 we obtain the relation

$$\mathbf{R}_{\lambda}\mathbf{w}^{o}(n) = \mathbf{p}_{\lambda} \quad (12),$$

which minimizes $J(n)$. We also find

$$\mathbf{R}_{\lambda} = \mathbf{X}^{\mathrm{T}}\Lambda\mathbf{X} = \sum_{\mathbf{k}=\mathbf{1}}^{\mathbf{n}} \lambda^{n-k}\mathbf{x}(k)\mathbf{x}^{\mathrm{T}}(k) \qquad (13)$$

$$\mathbf{p}_{\lambda} = \mathbf{X}^{\mathrm{T}}\Lambda\mathbf{d}(n) = \sum_{k=1}^{n} \lambda^{n-k}\mathbf{x}(k)d(k) \quad (14)$$

9.7.2:

Using (11), (13), and (14) of Problem 9.7.1 we find

$$\begin{aligned} J_{\min}(n) &= \mathbf{d}^{\mathrm{T}}(n)\Lambda\mathbf{d}(n) - 2\mathbf{d}^{\mathrm{T}}(n)\Lambda\mathbf{X}\hat{\mathbf{w}}(n) + \hat{\mathbf{w}}^{\mathrm{T}}(n)\mathbf{p}_{\lambda}(n) \\ &= \mathbf{d}^{\mathrm{T}}(n)\Lambda\mathbf{d}(n) - 2\hat{\mathbf{w}}^{\mathrm{T}}(n)\mathbf{X}^{\mathrm{T}}\Lambda\mathbf{d}(n) + \hat{\mathbf{w}}^{\mathrm{T}}(n)\mathbf{p}_{\lambda}(n) \\ &= \mathbf{d}^{\mathrm{T}}(n)\Lambda\mathbf{d}(n) - \hat{\mathbf{w}}^{\mathrm{T}}(n)\mathbf{p}_{\lambda}(n) \\ &= \mathbf{d}^{\mathrm{T}}(n)\Lambda\mathbf{d}(n) - \mathbf{p}_{\lambda}^{\mathrm{T}}(n)\hat{\mathbf{w}}(n). \end{aligned}$$

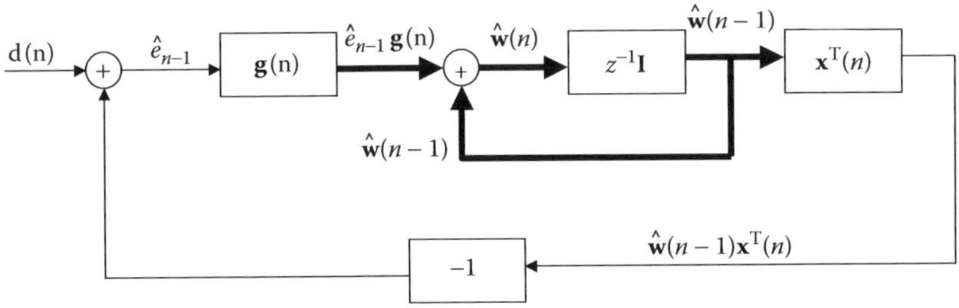

Figure p9.7.5 Solution of Problem 9.7.5.

9.7.3:

$\mathbf{A} + \mathbf{x}\mathbf{y}^T = \mathbf{A}(\mathbf{I} + \mathbf{A}^{-1}\mathbf{x}\mathbf{y})$ (1) and, hence, $(\mathbf{A} + \mathbf{x}\mathbf{y}^T)^{-1} = (\mathbf{I} + \mathbf{A}^{-1}\mathbf{x}\mathbf{y}^T)^{-1}\mathbf{A}^{-1}$ (2). Assuming $(\mathbf{I} + \mathbf{A})^{-1}$ exists, we find $(\mathbf{I} + \mathbf{A})^{-1} = \mathbf{I} - \mathbf{A} + A^2 - \mathbf{A}^3 + \ldots$ and hence $(\mathbf{I} + \mathbf{A}^{-1}\mathbf{x}\mathbf{y}^T)^{-1} = \mathbf{I} - \mathbf{A}^{-1}\mathbf{x}\mathbf{y}^T + (\mathbf{A}^{-1}\mathbf{x}\mathbf{y}^T)^2 - (\mathbf{A}^{-1}\mathbf{x}\mathbf{y}^T)^3 + \ldots = \mathbf{I} - \mathbf{A}^{-1}\mathbf{x}\mathbf{y}^T + \mathbf{A}^{-1}\mathbf{x}\mathbf{y}^T\mathbf{A}^{-1}\mathbf{x}\mathbf{y}^{T}\text{-}\ldots$ (3). Substituting (3) in (2) we obtain $(\mathbf{A} + \mathbf{x}\mathbf{y}^T)^{-1} = \mathbf{A}^{-1} - \mathbf{A}^{-1}\mathbf{x}\mathbf{y}^T\mathbf{A}^{-1} + \mathbf{A}^{-1}(\mathbf{y}^T\mathbf{A}^{-1}\mathbf{x})$

$$\mathbf{y}^T\mathbf{A}^{-1}\text{-} \ldots = \mathbf{A}^{-1} - \mathbf{A}^{-1}\mathbf{x}\mathbf{y}^T\mathbf{A}^{-1}(1 - \mathbf{y}^T\mathbf{A}^{-1}\mathbf{x} + (\mathbf{y}^T\mathbf{A}^{-1}\mathbf{x})^2 - \ldots) = \mathbf{A}^{-1} - \frac{\mathbf{A}^{-1}\mathbf{x}\mathbf{y}^T\mathbf{A}^{-1}}{1+\mathbf{y}^T\mathbf{A}^{-1}\mathbf{x}}.$$

9.7.4:

$\mathbf{A} = \mathbf{B}^{-1} + \mathbf{C}\mathbf{D}^{-1}\mathbf{C}^T$ (1); $\mathbf{A}^{-1} = \mathbf{B} - \mathbf{B}\mathbf{C}(\mathbf{D} + \mathbf{C}^T\mathbf{B}\mathbf{C})^{-1}\mathbf{C}^T\mathbf{B}$ (2). Multiply (1) by (2): $\mathbf{A}\mathbf{A}^{-1} = (\mathbf{B}^{-1} + \mathbf{C}\mathbf{D}^{-1}\mathbf{C}^T)[\mathbf{B} - \mathbf{B}\mathbf{C}(\mathbf{D} + \mathbf{C}^T\mathbf{B}\mathbf{C})^{-1}\mathbf{C}^T\mathbf{B}] = \mathbf{B}^{-1}\mathbf{B} - \mathbf{B}^{-1}\mathbf{B}\mathbf{C}\,()^{-1}\mathbf{C}^T\mathbf{B} + \mathbf{C}\mathbf{D}^{-1}\mathbf{C}^T\mathbf{B}\mathbf{C}()^{-1}\mathbf{C}^T\mathbf{B}$ (3),

or

$\mathbf{A}\mathbf{A}^{-1} = \mathbf{I} - \mathbf{C}()^{-1}\mathbf{C}^T\mathbf{B} + \mathbf{C}\mathbf{D}^{-1}()()^{-1}\mathbf{C}^T\mathbf{B} - \mathbf{C}\mathbf{D}^{-1}\mathbf{C}^T\mathbf{B}\mathbf{C}()^{-1}\mathbf{C}^T\mathbf{B} = \mathbf{I} - (\mathbf{C} - \mathbf{C}\mathbf{D}^{-1}(\mathbf{D} + \mathbf{C}^T\mathbf{B}\mathbf{C}) - \mathbf{C}\mathbf{D}^{-1}\mathbf{C}^T\mathbf{B}\mathbf{C})(\mathbf{D} + \mathbf{C}^T\mathbf{B}\mathbf{C})^{-1}\mathbf{C}^T\mathbf{B} = \mathbf{I} - (\mathbf{C} - \mathbf{C}\mathbf{D}^{-1}\mathbf{D})(\mathbf{D} + \mathbf{C}^T\mathbf{B}\mathbf{C})^{-1}\mathbf{C}^T\mathbf{B} = \mathbf{I}.$

9.7.5:

The diagram is shown in Figure p9.7.5.

Abbreviations

ANC	Adaptive Noise Cancellation
AR	Autoregressive
ARMA	Autoregressive Moving Average
BT	Blackman–Tukey
cdf	Cumulative Density Function
dB	Decibel
DFT	Discrete Fourier Transform
DTFT	Discrete-Time Fourier Transform
EDNSS	Error-Data Normalized Step-Size
EMSE	Excess Mean-Square Error
ENSS	Error Normalized Step-Size
fft	Fast Fourier Transform
FIR	Finite Impulse Response
IDFT	Inverse Discrete Fourier Transform
iid	Independent and Identically Distributed
IIR	Infinite Impulse Response
LMS	Least Mean-Square
LTI	Linear Time Invariant
MMSE	Minimum Mean-Square Error
MSE	Mean-Square Error
pdf	Probability Density Function
PSD	Power Spectral Density
QAM	Quadrature Amplitude Modulation
ROC	Region of Convergence
rv	Random Variable
RVSS	Robust Variable Step-Size
SCAF	Self-Correcting Adaptive Filter
SCWF	Self-Correcting Wiener Filter
SNR	Signal-to-Noise Ratio
WGN	White Gaussian Noise
WSS	Wide Sense Stationary

Bibliography

Brockwell, R.J. and Davis, R.A., *Time Series-Theory and Methods,* New York, Springer-Verlag, 1991.

Clarkson, P.M., *Optimal and Adaptive Signal Processing,* Boca Raton, FL, CRC Press, 1993.

Farhang-Boroujeny, B., *Adaptive Filters Theory and Applications,* New York, John Wiley & Sons, 1998.

Hayes, M.H., *Statistical Digital Signal Processing and Modeling,* New York, John Wiley & Sons, 1996.

Haykin, S., *Adaptive Filter Theory,* Upper Saddle River, NJ, Prentice Hall, 2001.

Haykin, S. and Widrow, B., *Least-Mean-Square Adaptive Filters,* Hoboken, NJ, Wiley Interscience, 2003.

Kay, S.M., *Fundamentals of Statistical Signal Processing-Estimation Theory,* Upper Saddle River, NJ, Prentice Hall, 1993.

Manolakis, D., Ingle V., Kogan S., and Kogan S.M., *Statistical and Adaptive Signal Processing: Spectral Estimation, Signal Modeling, Adaptive Filtering and Array Processing,* New York, McGraw-Hill, 1999.

Marple, L., *Digital Spectral Analysis with Applications,* Englewood Cliffs, NJ, Prentice Hall, 1987.

Sayed, A.H., *Fundamentals of Adaptive Filtering,* Hoboken, NJ, Wiley Interscience, 2004.

Stoica, P. and Moses, R., *Introduction to Spectra Analysis,* Upper Saddle River, NJ, Prentice Hall, 1997.

Widrow, B. and Stearns, S.D., *Adaptive Signal Processing,* Englewood Cliffs, NJ, Prentice Hall, 1985.

Appendix

Matrix analysis

A.1 Definitions

Let **A** be an $m \times n$ matrix with elements a_{ij}, i = 1, 2, … , m and j = 1, 2, … , n. A short hand description of **A** is

$$[\mathbf{A}]_{ij} = a_{ij} \tag{A.1}$$

The *transpose* of **A**, denoted by $\mathbf{A}^T$, is defined as the $n \times m$ matrix with elements a_{ji} or

$$\left[\mathbf{A}^T\right]_{ij} = a_{ji} \tag{A.2}$$

Example A.1.1:

$$\mathbf{A} = \begin{bmatrix} 1 & 2 \\ 4 & 9 \\ 3 & 1 \end{bmatrix}; \quad \mathbf{A}^T = \begin{bmatrix} 1 & 4 & 3 \\ 2 & 9 & 1 \end{bmatrix}$$

A *square* matrix is one for which $m = n$. A square matrix is symmetric if $\mathbf{A}^T = \mathbf{A}$.

The *rank* of a matrix is the number of linearly independent rows or columns, whichever is less. The *inverse* of a square $n \times n$ matrix $\mathbf{A}^{-1}$ for which

$$\mathbf{A}^{-1}\mathbf{A} = \mathbf{A}\mathbf{A}^{-1} = \mathbf{I} \tag{A.3}$$

where **I** is the identity matrix

$$\mathbf{I} = \begin{bmatrix} 1 & 0 & \cdots & 0 \\ 0 & 1 & \cdots & 0 \\ \vdots & \vdots & & \vdots \\ 0 & 0 & \cdots & 1 \end{bmatrix} \tag{A.4}$$

A matrix **A** is singular if its inverse does not exist.

The *determinant* of a square $n \times n$ matrix, **A**, is denoted by det{**A**}, and it is computed as

$$\det\{\mathbf{A}\} = \sum_{j=1}^{n} a_{ij} C_{ij} \tag{A.5}$$

where

$$C_{ij} = (-1)^{i+j} M_{ij} \tag{A.6}$$

and M_{ij} is the determinant of the sub matrix **A** is obtained by deleting the ith row and j^{th} column and is called the minor of a_{ij}. C_{ij} is the *cofactor* of a_{ij}.

Example A.1.2:

$$\mathbf{A} = \begin{bmatrix} 1 & 2 & 4 \\ 4 & -3 & 9 \\ -1 & -1 & 6 \end{bmatrix}, \quad \det\{A\} = (-1)^{1+1} \begin{vmatrix} -3 & 9 \\ -1 & 6 \end{vmatrix} + (-1)^{1+2} 2 \begin{vmatrix} 4 & 9 \\ -1 & 6 \end{vmatrix}$$

$$+ (-1)^{1+3} 4 \begin{vmatrix} 4 & -3 \\ -1 & -1 \end{vmatrix} = C_{11} + C_{12} + C_{13} = (18+9) + [-2(24+9)] + [4(-4-3)]$$

Any choice of i or j will yield the same value for the det{**A**}.

A *quadratic form* Q associated with an $n \times n$ matrix **A** is defined by

$$Q = \sum_{n=1}^{n} \sum_{j=1}^{n} a_{ij} x_i x_j \tag{A.7}$$

In defining the quadratic form, it is assumed that $a_{ji} = a_{ij}$. This entails no loss in generality since any quadratic functions may be expressed in this manner. Q may also be expressed as

$$Q = \mathbf{x}^T \mathbf{A} \mathbf{x} \tag{A.8}$$

where $\mathbf{x} = [x_1 \ \ x_2 \ \ldots \ x_n]^T$ and $\mathbf{A}$ is a square $n \times n$ matrix with $a_{ji} = a_{ij}$ (symmetric matrix).
Note that Q is just the inner product of $\mathbf{x}$ and $\mathbf{Ax}$. That is, $Q = \langle \mathbf{x}, \mathbf{Ax} \rangle$.

Example A.1.3:

$$Q = \begin{bmatrix} x_1 & x_2 \end{bmatrix} \begin{bmatrix} a_{11} & a_{12} \\ a_{21} & a_{22} \end{bmatrix} \begin{bmatrix} x_1 \\ x_2 \end{bmatrix} = [a_{11}x_1 + a_{21}x_2 \quad a_{12}x_1 + a_{22}x_2] \begin{bmatrix} x_1 \\ x_2 \end{bmatrix}$$

$$= a_{11}x_1^2 + a_{21}x_1x_2 + a_{12}x_1x_2 + a_{22}x_2^2.$$

A square $n \times n$ matrix $\mathbf{A}$ is positive *semi-definite* if $\mathbf{A}$ is symmetric and

$$\mathbf{x}^T \mathbf{A} \mathbf{x} \geq 0 \tag{A.9}$$

for all $x \neq 0$. If the quadratic form is strictly positive, the matrix $\mathbf{A}$ is called *positive definite*. If a matrix is positive definite or positive semi-definite it is automatically assumed that the matrix is symmetric.

The *trace* of a square matrix is the sum of the diagonal elements or

$$tr\{\mathbf{A}\} = \sum_{i=1}^{n} a_{ii} \tag{A.10}$$

A partitioned $m \times n$ matrix $\mathbf{A}$ is one that is expressed in terms of its sub-matrices. An example is the 2×2 partitioned

$$\mathbf{A} - \begin{bmatrix} \mathbf{A}_{11} & \mathbf{A}_{12} \\ \mathbf{A}_{21} & \mathbf{A}_{22} \end{bmatrix}, \text{ with corresponding dimensions } \begin{bmatrix} k \times l & k \times (n-l) \\ (m-k) \times l & (m-k) \times (n-l) \end{bmatrix} \tag{A.11}$$

MATLAB functions

```
B=A';% B is the transpose of A
B=inv(A);%B is the inverse of A
a=det(A);% a is the determinant of A
I=eye(n);%I is an nxn identity matrix
a=trace(A);%a is the trace of A
```

A.2 Special matrices

A *diagonal matrix* is a square $n \times n$ matrix with $a_{ij} = 0$ for $i \neq j$. A diagonal matrix has all the elements of the principal diagonal equal to zero. Hence

$$\mathbf{A} = \begin{bmatrix} a_{11} & 0 & \cdots & 0 \\ 0 & a_{22} & \cdots & 0 \\ \vdots & \vdots & & \vdots \\ 0 & 0 & \cdots & a_{nn} \end{bmatrix} \tag{A.12}$$

$$\mathbf{A}^{-1} = \begin{bmatrix} a_{11}^{-1} & 0 & \cdots & 0 \\ 0 & a_{22}^{-1} & \cdots & 0 \\ \vdots & \vdots & & \vdots \\ 0 & 0 & \cdots & a_{nn}^{-1} \end{bmatrix} \tag{A.13}$$

A generalization of the diagonal matrix is the square $n \times n$ block diagonal matrix

$$\mathbf{A} = \begin{bmatrix} \mathbf{A}_{11} & 0 & \cdots & 0 \\ 0 & \mathbf{A}_{22} & \cdots & 0 \\ \vdots & \vdots & & \vdots \\ 0 & 0 & \cdots & \mathbf{A}_{kk} \end{bmatrix} \tag{A.14}$$

where all $\mathbf{A}_{ii}$ matrices are square and the submatrices are identically zero. The submatrices may not have the same dimensions. For example, if $k = 2$, A_{11} might be a 2×2 matrix, and A_{22} might be a scalar. If all $\mathbf{A}_{ii}$ are nonsingular, then

$$\mathbf{A}^{-1} = \begin{bmatrix} \mathbf{A}_{11}^{-1} & 0 & \cdots & 0 \\ 0 & \mathbf{A}_{22}^{-1} & \cdots & 0 \\ \vdots & \vdots & & \vdots \\ 0 & 0 & \cdots & \mathbf{A}_{kk}^{-1} \end{bmatrix} \tag{A.15}$$

and

$$\det\{\mathbf{A}\} = \prod_{i=1}^{n} \det\{\mathbf{A}_{ii}\} \tag{A.16}$$

A square $n \times n$ matrix is *orthogonal* if

$$\mathbf{A}^{-1} = \mathbf{A}^{\mathrm{T}} \tag{A.17}$$

Example A.2.1:

$$\mathbf{A} = \begin{bmatrix} \frac{2}{\sqrt{5}} & \frac{1}{\sqrt{5}} \\ -\frac{1}{\sqrt{5}} & \frac{2}{\sqrt{5}} \end{bmatrix}, \quad \mathbf{A}^{-1} = \frac{1}{\det\{\mathbf{A}\}} \begin{bmatrix} \frac{2}{\sqrt{5}} & \frac{1}{\sqrt{5}} \\ -\frac{1}{\sqrt{5}} & \frac{2}{\sqrt{5}} \end{bmatrix}^{\mathrm{T}} = \begin{bmatrix} \frac{2}{\sqrt{5}} & -\frac{1}{\sqrt{5}} \\ \frac{1}{\sqrt{5}} & \frac{2}{\sqrt{5}} \end{bmatrix} = \mathbf{A}^{\mathrm{T}}$$

A matrix is *orthogonal* if its columns (and rows) are *orthonormal*. Therefore, we must have

$$\mathbf{A} = \begin{bmatrix} \mathbf{a}_1 & \mathbf{a}_2 & \cdots & \mathbf{a}_{\mathrm{n}} \end{bmatrix}$$
$$\mathbf{a}_i^{\mathrm{T}} \mathbf{a}_j = \begin{Bmatrix} 0 & \mathit{for} & i \neq j \\ 1 & \mathit{for} & i = j \end{Bmatrix} \tag{A.18}$$

An *idempotent* matrix is a square $n \times n$ matrix that satisfies the relations

$$A^2 = A$$
$$A^m = A \tag{A.19}$$

Example A.2.2:
The *projection* matrix $A = H(H^TH)^{-1}H^T$ becomes $A^2 = H(H^TH)^{-1}H^TH(H^TH)^{-1}H^T = H(H^{-1}H^{-T}H^TH(H^TH)^{-1})H^T = H(H^{-1}IH(H^TH)^{-1})H^T = H(H^TH)^{-1}H^T$, and hence, it is an idempotent matrix.

A *Toeplitz* square matrix is defined as

$$[\mathrm{A}]_{ij} = a_{i-j} \tag{A.20}$$

$$\mathbf{A} = \begin{bmatrix} a_0 & a_{-1} & a_{-2} & \cdots & a_{-(n-1)} \\ a_1 & a_0 & a_{-1} & \cdots & a_{-(n-2)} \\ \vdots & \vdots & \vdots & & \vdots \\ a_{n-1} & a_{n-2} & a_{n-3} & \cdots & a_0 \end{bmatrix} \tag{A.21}$$

Each element along the northwest-to-southeast diagonals is the same. If, in addition, $a_{-k} = a_k$, then **A** is *symmetric Toeplitz.*

MATLAB functions

```
A=diag(x);%creates a diagonal matrix A with its diagonal
%the vector x;
A=toeplitz(x);%A is a symmetric Toeplitz matrix;
A=toeplitz(x,y) % x, and y must be of the same length,
%the first element of the main
%diagonal will be the first
%element of x, the first element of y is not used;
```

A.3 Matrix operation and formulas

Addition and Subtraction

$$\mathbf{A}+\mathbf{B}=\begin{bmatrix} a_{11}\ a_{12}\cdots a_{1n} \\ a_{21}\ a_{22}\cdots a_{2n} \\ \vdots \\ a_{m1}\ a_{m2}\cdots a_{mn} \end{bmatrix}+\begin{bmatrix} b_{11}\ b_{12}\cdots b_{1n} \\ b_{21}\ b_{22}\cdots b_{2n} \\ \vdots \\ b_{m1}\ b_{m2}\cdots b_{mn} \end{bmatrix}=\begin{bmatrix} a_{11}+b_{11}\ a_{12}+b_{12}\cdots\ a_{1n}+b_{1n} \\ a_{21}+b_{21}\ a_{22}+b_{22}\cdots\ a_{2n}+b_{2n} \\ \vdots \\ a_{m1}+b_{m1}\ a_{m2}+b_{m2}\cdots a_{mn}+b_{mn} \end{bmatrix} \tag{A.22}$$

Both matrices must have the same dimension.

Multiplication

$$\mathbf{AB}\ (m\times n\times n\times k)=\mathbf{C}(m\times k)$$

$$c_{ij}=\sum_{j=1}^{n} a_{ij}b_{ji} \tag{A.23}$$

Example A.3.1:

$$\mathbf{AB}=\begin{bmatrix} a_{11} & a_{12} \\ a_{21} & a_{22} \\ a_{31} & a_{32} \end{bmatrix}\begin{bmatrix} b_{11} & b_{12} \\ b_{21} & b_{22} \end{bmatrix}=\begin{bmatrix} a_{11}b_{11}+a_{12}b_{21} & a_{11}b_{12}+a_{12}b_{22} \\ a_{21}b_{11}+a_{22}b_{21} & a_{21}b_{12}+a_{22}b_{22} \\ a_{31}b_{11}+a_{32}b_{21} & a_{31}b_{12}+a_{32}b_{22} \end{bmatrix} \tag{A.24}$$

$$3\times 2\times 2\times 2=3\times 2$$

Transposition

$$(\mathbf{AB})^{\mathrm{T}}=\mathbf{B}^{\mathrm{T}}\mathbf{A}^{\mathrm{T}} \tag{A.25}$$

Inversion

$$(\mathbf{A}^{\mathrm{T}})^{-1} = (\mathbf{A}^{-1})^{\mathrm{T}} \tag{A.26}$$

$$(\mathbf{AB})^{-1} = \mathbf{B}^{-1}\mathbf{A}^{-1} \tag{A.27}$$

$$\mathbf{A}^{-1} = \frac{\mathbf{C}^T}{\det\{\mathbf{A}\}} \qquad (\mathbf{A} \equiv n \times n \ \ matrix) \tag{A.28}$$

$$c_{ij} = (-1)^{i+j} M_{ij} \tag{A.29}$$

$M_{ij} \equiv$ minor of a_{ij} obtained by deleting the ith row and j^{th} column of $\mathbf{A}$

Example A.3.2:

$$\mathbf{A}^{-1} = \begin{bmatrix} 2 & 4 \\ -1 & 5 \end{bmatrix}^{-1} = \frac{1}{10+4}\begin{bmatrix} 5 & 1 \\ -4 & 2 \end{bmatrix}^{T} = \frac{1}{14}\begin{bmatrix} 5 & -4 \\ 1 & 2 \end{bmatrix}; \ AA^{-1} = \begin{bmatrix} 2 & 4 \\ -1 & 5 \end{bmatrix}\frac{1}{14}\begin{bmatrix} 5 & -4 \\ 2 & 2 \end{bmatrix}$$

$$= \frac{1}{14}\begin{bmatrix} 14 & -8+8 \\ -5+5 & 4+10 \end{bmatrix} = \begin{bmatrix} 1 & 0 \\ 0 & 1 \end{bmatrix} = \mathbf{I}$$

Determinant (see (A.5))

$\mathbf{A} = n \times n$ *matrix*; $\mathbf{B} = n \times n$ *matrix*

$$\det\{\mathbf{A}^{\mathrm{T}}\} = \det\{\mathbf{A}\} \tag{A.30}$$

$$\det\{c\mathbf{A}\} = c^n \det\{\mathbf{A}\} \tag{A.31}$$

$$\det\{\mathbf{AB}\} = \det\{\mathbf{A}\}\det\{\mathbf{B}\} \tag{A.32}$$

$$\det\{\mathbf{A}^{-1}\} = \frac{1}{\det\{\mathbf{A}\}} \tag{A.33}$$

Trace (see (A.10))

$\mathbf{A} = n \times n$ *matrix*; $\mathbf{B} = n \times n$ *matrix*

$$tr\{\mathbf{AB}\} = tr\{\mathbf{BA}\} \tag{A.34}$$

$$tr\{\mathbf{A}^{\mathrm{T}}\mathbf{B}\} = \sum_{i=1}^{n}\sum_{j=1}^{n} a_{ij} b_{ij} \tag{A.35}$$

$$tr\{\mathbf{x}\mathbf{y}^{\mathrm{T}}\} = \mathbf{y}^{\mathrm{T}}\mathbf{x}; \quad \mathbf{x}, \mathbf{y} = vectors \tag{A.36}$$

Matrix inversion formula

$\mathbf{A} = n\times n,\ \mathbf{B} = n\times m,\ \mathbf{C} = m\times m,\ \mathbf{D} = m\times n$

$$(\mathbf{A}+\mathbf{BCD})^{-1} = \mathbf{A}^{-1} - \mathbf{A}^{-1}\mathbf{B}(\mathbf{DA}^{-1}\mathbf{B}+\mathbf{C}^{-1})^{-1}\mathbf{DA}^{-1} \tag{A.37}$$

$$(\mathbf{A}+\mathbf{xx}^{\mathrm{T}})^{-1} = \mathbf{A}^{-1} - \frac{\mathbf{A}^{-1}\mathbf{xx}^{\mathrm{T}}\mathbf{A}^{-1}}{1+\mathbf{x}^{\mathrm{T}}\mathbf{A}^{-1}\mathbf{x}},\quad \mathbf{x} = n\times 1\ \textit{vector} \tag{A.38}$$

Partition matrices

Examples of 2 × 2 partition matrices are given below.

$$\mathbf{AB} = \begin{bmatrix} \mathbf{A}_{11} & \mathbf{A}_{12} \\ \mathbf{A}_{21} & \mathbf{A}_{22} \end{bmatrix}\begin{bmatrix} \mathbf{B}_{11} & \mathbf{B}_{12} \\ \mathbf{B}_{21} & \mathbf{B}_{22} \end{bmatrix} = \begin{bmatrix} \mathbf{A}_{11}\mathbf{B}_{11}+\mathbf{A}_{12}\mathbf{B}_{21} & \mathbf{A}_{11}\mathbf{B}_{2}+\mathbf{A}_{12}\mathbf{B}_{22} \\ \mathbf{A}_{21}\mathbf{B}_{11}+\mathbf{A}_{22}\mathbf{B}_{21} & \mathbf{A}_{21}\mathbf{B}_{12}+\mathbf{A}_{22}\mathbf{B}_{22} \end{bmatrix} \tag{A.39}$$

$$\begin{bmatrix} \mathbf{A}_{11} & \mathbf{A}_{12} \\ \mathbf{A}_{21} & \mathbf{A}_{22} \end{bmatrix}^{\mathrm{T}} = \begin{bmatrix} \mathbf{A}_{11}^{\mathrm{T}} & \mathbf{A}_{12}^{\mathrm{T}} \\ \mathbf{A}_{21}^{\mathrm{T}} & \mathbf{A}_{22}^{\mathrm{T}} \end{bmatrix} \tag{A.40}$$

$$\mathbf{A} = \begin{bmatrix} \mathbf{A}_{11} & \mathbf{A}_{12} \\ \mathbf{A}_{21} & \mathbf{A}_{22} \end{bmatrix} = \begin{bmatrix} k\times k & k\times(n-k) \\ (n-k)\times k & (n-k)\times(n-k) \end{bmatrix}$$

$$\mathbf{A}^{-1} = \begin{bmatrix} (\mathbf{A}_{11}-(\mathbf{A}_{12}\mathbf{A}_{22}^{-1}\mathbf{A}_{21})^{-1} & -(\mathbf{A}_{11}-\mathbf{A}_{12}\mathbf{A}_{22}^{-1}\mathbf{A}_{21})^{-1}\mathbf{A}_{12}\mathbf{A}_{22}^{-1} \\ -(\mathbf{A}_{22}-\mathbf{A}_{21}\mathbf{A}_{11}^{-1}\mathbf{A}_{12})^{-1}\mathbf{A}_{21}\mathbf{A}_{11}^{-1} & (\mathbf{A}_{22}-\mathbf{A}_{21}\mathbf{A}_{11}^{-1}\mathbf{A}_{12})^{-1} \end{bmatrix} \tag{A.41}$$

$$\det\{\mathbf{A}\} = \det\{\mathbf{A}_{12}\}\det\{\mathbf{A}_{11}-\mathbf{A}_{12}\mathbf{A}_{22}^{-1}\mathbf{A}_{21}\} = \det\{\mathbf{A}_{11}\}\det\{\mathbf{A}_{22}-\mathbf{A}_{21}\mathbf{A}_{11}^{-1}\mathbf{A}_{12}\} \tag{A.42}$$

Important Theorems

1. A square matrix **A** is singular (invertible) if and only if its columns (or rows) are linearly independent or, equivalently, if $\det\{\mathbf{A}\} \neq 0$. If this is true, **A** is called a *full rank* matrix. Otherwise, it is singular.
2. A square matrix **A** is positive definite if and only if
   a. $\mathbf{A} = \mathbf{CC}^{\mathrm{T}}$ (A.43)
   where **C** is a square matrix of the same dimension as **A** and it is of full rank (invertible), or
   b. the principal minors are all positive. (The ith principal minor is the determinant of the submatrix formed by deleting all rows and

columns with an index greater than i). If **A** can be written as in (A.43), and **C** is not full rank or the principal minor are only nonnegative, then **A** is positive definite.

3. If **A** is positive definite, then

$$\mathbf{A}^{-1} = (\mathbf{C}^{-1})^{\mathrm{T}}\mathbf{C}^{-1} \tag{A.44}$$

4. If **A** is positive definite and **B** $(m \times n)$ is of full rank $(m \le n)$, then $\mathbf{BAB}^{\mathrm{T}}$ is positive definite.
5. If **A** is positive definite (or positive semi-definite), then the diagonal elements are positive (nonnegative).

A.4 Eigendecomposition of matrices

Let λ denote an *eigenvalue* of the matrix **A** $(n \times n)$, then

$$\mathbf{Av} = \lambda\mathbf{v} \tag{A.45}$$

where **v** is the *eigenvector* corresponding to the eigenvalue λ. If **A** is symmetric, then

$$\mathbf{Av}_i = \lambda_i\mathbf{v}_i,\ \mathbf{Av}_j = \lambda_j\mathbf{v}_j \quad (\lambda_i \ne \lambda_j) \text{ and}$$

$$\mathbf{v}_j^{\mathrm{T}}\mathbf{Av}_i = \lambda_i\mathbf{v}_j^{\mathrm{T}}\mathbf{v}_i \qquad \text{(a)}$$

$$\mathbf{v}_i^{\mathrm{T}}\mathbf{Av}_j = \lambda_j\mathbf{v}_i^{\mathrm{T}}\mathbf{v}_j \quad or \quad \mathbf{v}_j^{\mathrm{T}}\mathbf{Av}_i = \lambda_j\mathbf{v}_j^{\mathrm{T}}\mathbf{v}_i \qquad \text{(b)}$$

Subtracting (a) from (b) we obtain $(\lambda_i - \lambda_j)\mathbf{v}_j^{\mathrm{T}}\mathbf{v}_i = 0$. But $\lambda_i \ne \lambda_j$ and hence $\mathbf{v}_j^{\mathrm{T}}\mathbf{v}_i = 0$, which implies that the eigenvectors of a symmetric matrix are orthogonal. We can proceed and normalize them producing orthonormal eigenvectors.

From (A.45) we write

$$\mathbf{A}[\mathbf{v}_1 \ \mathbf{v}_2 \ \cdots \ \mathbf{v}_n] = [\lambda_1\mathbf{v}_1 \ \ \lambda_2\mathbf{v}_2 \ \cdots \ \lambda_n\mathbf{v}_n]$$

or (A.46)

$$\mathbf{AV} = \mathbf{V}\Lambda$$

where

$$\Lambda = \begin{bmatrix} \lambda_1 & 0 & 0 \cdots 0 \\ 0 & \lambda_2 & 0 \cdots 0 \\ & \vdots & \\ 0 & 0 & 0 \cdots \lambda_n \end{bmatrix} \tag{A.47}$$

and **V** is an $n \times n$ matrix whose columns are the eigenvectors of **A**.

Because $\mathbf{v}_i$ are mutually orthogonal, $\mathbf{v}_i^T\mathbf{v}_j = \delta_{ij}$ makes $\mathbf{V}$ a *unitary* matrix, $\mathbf{V}^T\mathbf{V} = \mathbf{I} = \mathbf{V}\mathbf{V}^T$. Post-multiply (A.46) by $\mathbf{V}^T$, we obtain

$$\mathbf{A} = \mathbf{V}\Lambda\mathbf{V}^T = \sum_{i=1}^{n} \lambda_i \mathbf{v}_i \mathbf{v}_i^T \tag{A.48}$$

which is known as *unitary decomposition* of $\mathbf{A}$. We also say that $\mathbf{A}$ is *unitary similar* to the diagonal , because a unitary matrix $\mathbf{V}$ takes $\mathbf{A}$ to diagonal form: $\mathbf{V}^T\mathbf{A}\mathbf{V} = \Lambda$.

If $\Lambda = \mathbf{I}$, then from (A.48) $\mathbf{A} = \mathbf{V}\mathbf{V}^T = \mathbf{I}$ and, hence,

$$\mathbf{I} = \mathbf{V}\mathbf{V}^T = \sum_{i=1}^{n} \mathbf{v}_i \mathbf{v}_i^T \tag{A.49}$$

Each of the terms in the summation is of rank 1 projection matrix:

$$\mathbf{P}_i^2 = \mathbf{v}_i\mathbf{v}_i^T\mathbf{v}_i\mathbf{v}_i^T = \mathbf{v}_i\mathbf{v}_i^T = \mathbf{P}_i \qquad (\mathbf{v}_i^T \mathrm{v}_i = 1) \tag{A.50}$$

$$\mathbf{P}_i^T = \mathbf{v}_i\mathbf{v}_i^T = \mathbf{P}_i \tag{A.51}$$

Hence, we write (see (A.48) and (A.49))

$$\mathbf{A} = \sum_{i=1}^{n} \lambda_i \mathbf{P}_i \tag{A.52}$$

$$\mathbf{I} = \sum_{i=1}^{n} \mathbf{P}_i \tag{A.53}$$

Inverse

Because $\mathbf{V}$ is unitary matrix, $\mathbf{V}\mathbf{V}^T = \mathbf{I}$ or $\mathbf{V}^T = \mathbf{V}^{-1}$ or $\mathbf{V} = (\mathbf{V}^T)^{-1}$ and, therefore,

$$\mathbf{A}^{-1} = (\mathbf{V}^T)^{-1}\Lambda^{-1}\mathbf{V}^{-1} = \mathbf{V}\Lambda^{-1}\mathbf{V}^T = \sum_{i=1}^{n} \frac{1}{\lambda_i} v_i v_i^T \tag{A.54}$$

Determinant

$$\det\{\mathbf{A}\} = \det\{\mathbf{V}\}\det\{\Lambda\}\det\{\mathbf{V}^{-1}\} = \det\{\Lambda\} = \prod_{i=1}^{n} \lambda_i \tag{A.55}$$

A.5 Matrix expectations

Let **x** be a random variable, **A** be a matrix, and **a** and **b** be vectors of appropriate dimensions, and assume that all the elements of **A**, **a**, and **b** are deterministic. Then

$$E\{\mathbf{x}\} = m_x \tag{A.56}$$

$$E\{(\mathbf{x} - m_x)(\mathbf{x} - m_x)^{\mathrm{T}}\} = \mathbf{R}_{xx} \tag{A.57}$$

$$E\{tr(\mathbf{A})\} = tr(E\{\mathbf{A}\}) \tag{A.58}$$

$$E\{\mathbf{Ax} + \mathbf{b}\} = \mathbf{A}m_x + \mathbf{b} \tag{A.59}$$

$$E\{\mathbf{xx}^{\mathrm{T}}\} = \mathbf{R}_{xx} + m_x m_x^{\mathrm{T}} \tag{A.60}$$

$$E\{\mathbf{xa}^{\mathrm{T}}\mathbf{x}\} = (\mathbf{R}_{xx} + m_x m_x^{\mathrm{T}})\mathbf{a} \tag{A.61}$$

$$E\{(\mathrm{x} + \mathrm{a})(\mathrm{x} + \mathbf{a})^{\mathrm{T}}\} = \mathbf{R}_{xx} + (m_x + \mathbf{a})(m_x + \mathrm{a})^{\mathrm{T}} \tag{A.62}$$

$$E\{\mathbf{x}^{\mathrm{T}}\mathbf{x}\} = tr\{\mathbf{R}_{xx}\} + m_x^{\mathrm{T}} m_x = tr\{\mathbf{R}_{xx} + m_x m_x^{\mathrm{T}}\} \tag{A.63}$$

$$E\{\mathbf{x}^{\mathrm{T}}\mathbf{ax}^{\mathrm{T}}\} = \mathbf{a}^{\mathrm{T}}[\mathbf{R}_{xx} + m'_x m_x^{\mathrm{T}}] \tag{A.64}$$

$$E\{\mathbf{x}^{\mathrm{T}}\mathbf{Ax}\} = tr\{\mathbf{AR}_{xx}\} + m_x^{\mathrm{T}}\mathbf{A}m_x = tr\{\mathbf{A}(\mathbf{R}_{xx} + m_x m_x^{\mathrm{T}})\} \tag{A.65}$$

A.6 Differentiation of a scalar function with respect to a vector

Let **x** and **y** be vector columns, and **A** be a matrix of appropriate dimensions. Assume that $\mathbf{x} = [x_1 \quad x_2 \quad \cdots \quad x_n]^{\mathrm{T}}$, then $\frac{\partial}{\partial \mathbf{x}} = [\frac{\partial}{\partial \mathbf{x_1}} \quad \frac{\partial}{\partial \mathbf{x_2}} \quad \cdots \quad \frac{\partial}{\partial \mathbf{x_n}}]^{\mathrm{T}}$ and

$$\frac{\partial}{\partial \mathbf{x}}(\mathbf{y}^{\mathrm{T}}\mathbf{x}) = \frac{\partial}{\partial \mathbf{x}}(\mathbf{x}^{\mathrm{T}}\mathbf{y}) = \mathbf{y} \tag{A.66}$$

$$\frac{\partial}{\partial \mathbf{x}}(\mathbf{x}^{\mathrm{T}}\mathbf{A}) = \mathbf{A} \tag{A.67}$$

$$\frac{\partial}{\partial \mathbf{x}}(\mathbf{x}^{\mathrm{T}}) = \mathbf{I} \tag{A.68}$$

$$\frac{\partial}{\partial \mathbf{x}}(\mathbf{x}^{\mathrm{T}}\mathbf{x}) = 2\mathbf{x} \tag{A.69}$$

$$\frac{\partial}{\partial \mathbf{x}}(\mathbf{x}^{\mathrm{T}}\mathbf{A}\mathbf{y}) = \mathbf{A}\mathbf{y} \tag{A.70}$$

$$\frac{\partial}{\partial \mathbf{x}}(\mathbf{y}^{\mathrm{T}}\mathbf{A}\mathbf{x}) = \mathbf{A}^{\mathrm{T}}\mathbf{y} \tag{A.71}$$

$$\frac{\partial}{\partial \mathbf{x}}(\mathbf{x}^{\mathrm{T}}\mathbf{A}\mathbf{x}) = (\mathbf{A} + \mathbf{A}^{\mathrm{T}})\mathbf{x} \tag{A.72}$$

$$\frac{\partial}{\partial \mathbf{x}}(\mathbf{x}^{\mathrm{T}}\mathbf{A}\mathbf{x}) = 2\mathbf{A}\mathbf{x} \quad \text{if } \mathbf{A} \text{ is symmetric} \tag{A.73}$$

$$\frac{\partial}{\partial \mathbf{x}}[\mathbf{a}^{\mathrm{T}}\mathbf{A}\mathbf{x}\mathbf{x}^{\mathrm{T}}) = (\mathbf{A} + \mathbf{A}^{\mathrm{T}})\mathbf{x}\mathbf{x}^{\mathrm{T}} + \mathbf{x}^{\mathrm{T}}\mathbf{A}\mathbf{x}\mathbf{I} \tag{A.74}$$

Index

A

Autocorrelation matrix, 27
  example, 27–29
    solution, 27–29
Autoregressive moving average process, 36–38
Autoregressive process, 38
  example, 38
    solution, 38–39

B

Bartlett method, nonparametric spectra estimation, 44
Blackman–Tukey method, nonparametric spectra estimation, 42–44
  Bartlett (triangle) window, 43
  Blackman window, 43
  Hamming window, 43
  Hann window, 43
  Kaiser window, 43
  rectangle window, 42
Blackman–Tukey periodogram with Bartlett window, nonparametric spectra estimation, 48

D

Discrete Fourier transform, 8
  example, 8–9
    solution, 9–10
Discrete-time Fourier transform, 6–11
  example, 6–7
    solution, 7
Discrete-time signal processing, 5–18
  discrete-time signals, 5–6
    basic discrete-time signals, 5–6
  discrete-time systems, 13–18
    example, 13
      solution, 13
    hints-solutions-suggestions, 17–18
    problems, 17
    transform-domain representation, 14–16
  transform-domain representation of discrete-time signals, 6–11
    discrete Fourier transform, 8
      example, 8–9
        solution, 9–10
    discrete-time Fourier transform, 6–11
      example, 6–7
  z-transform, 11–12
    example, 11
      solution, 11–12
Discrete-time systems, 13–18
  example, 13
    solution, 13
  hints-solutions-suggestions, 17–18
  problems, 17
  transform-domain representation, 14–16

E

Eigenvalues
  error surface properties, 77–84
  properties of error surface, 77–84
    correlation matrix, 77–79
    geometrical properties, 79–84
      example, 80–81
      hints-solutions-suggestions, 82–84
      problems, 81–82
Ergodic processes, 21–22
Error normalized least mean-square algorithms, 158–169
  hints-solutions-suggestions, 167–169
  problems, 167
  simulations, 159–166

Error surface, eigenvalues, 77–84
correlation matrix, eigenvalues of, 77–79
geometrical properties, 79–84
example, 80–81
hints-solutions-suggestions, 82–84
problems, 81–82
Example, 35
solution, 35

F

Filtering random processes, 34–36
example, 35
solution, 35
spectral factorization, 35–36
Finite impulse response Wiener filter, 55–59
example, 57
solution, 57–58

G

Gradient search, Newton method, 85–99
one-dimensional gradient search method, 85–91
gradient search algorithm, 86–87
Newton method in gradient search, 87–88
Newton multidimensional case, 88–91

L

Leaky least mean-square algorithm, 142–145
Least mean-square algorithm, 101–135
complex representation of, 126–135
example, 129
hints-solutions-suggestions, 130–135
problems, 129–130
derivation of, 102–104
examples, 104–112
inverse system identification, 112
linear prediction, 105
modeling, 106–107
noise cancellation, 107
power spectrum approximation, 107
performance analysis
misadjustment, 119–121
steepest-descent method, 122–125
performance analysis of, 112–125
coefficient-error, weighted-error correlation matrix, 117–119
excess mean-square error, 119–121
learning curve, 114–117
example, 116–117
least mean-square method, 122–125
channel equalization example, 122–125
stability, 121–122
Least mean-square algorithms, variations, 137–169
error normalized least mean-square algorithms, 158–169
hints-solutions-suggestions, 167–169
problems, 167
simulations, 159–166
leaky least mean-square algorithm, 142–145
linearly constrained least mean-square algorithm, 145–150
normalized least mean-square algorithm, 139–141
self-correcting adaptive filtering, 150–153
sign algorithms, 137–139
error sign algorithm, 137–138
normalized least mean-square sign algorithm, 138
sign-sign algorithm, 139
signed-regressor algorithm, 139
summary, 166
transform domain adaptive least mean-square filtering, 153–157
convergence, 156–157
example, 155–156
variable step-size least mean-square algorithm, 141–142
Least-squares finite impulse response filter, 186–188
example, 187–188
Least squares signal processing, 171–201
least-square formulation, 171–180
statistical properties of least-squares estimators, 177
least-squares approach, 180–182
example, 180–182
least-squares finite impulse response filter, 186–188
example, 187–188
orthogonality principle, 182–183
corollary, 183
example, 183
projection operator, 184–185
example, 185
recursive least-square algorithm, 188–201
example, 194
hints-solutions-suggestions, 197–201
initialization of, 193–194
matrix inversion lemma, 191
problems, 197–201
summary, 194–196
Linearly constrained least mean-square algorithm, 145–150

M

Moving average process, 40

N

Newton method, gradient search, 85–99
  one-dimensional gradient search method, 85–91
    gradient search algorithm, 86–87
    Newton method in gradient search, 87–88
    Newton multidimensional case, 88–91
Nonparametric spectra estimation, 40–49
  Bartlett method, 44
  Blackman–Tukey method, 42–44
    Bartlett (triangle) window, 43
    Blackman window, 43
    Hamming window, 43
    Hann window, 43
    Kaiser window, 43
    rectangle window, 42
  Blackman–Tukey periodogram with Bartlett window, 48
  correlogram, 40
  modified Welch method, 46
  periodogram, 40
  Welch method, 45–46
Normalized least mean-square algorithm, 139–141

O

One-dimensional gradient search method, 85–91
  gradient search algorithm, 86–87
  Newton method in gradient search, 87–88
  Newton multidimensional case, 88–91
Orthogonality condition, 62
Orthogonality principle, 182–183
  corollary, 183
  example, 183

P

Parametric methods, power spectral estimations, 49–54
  autoregressive process, 50–51
  hints-solutions-suggestions, 52–54
  moving average process, 50
  problems, 51–52
Parametric methods of power spectral estimations, 49–54
  autoregressive process, 50–51
  hints-solutions-suggestions, 52–54
  moving average process, 50
  problems, 51–52
Power spectral estimations, 49–54
  autoregressive process, 50–51
  hints-solutions-suggestions, 52–54
  moving average process, 50
  problems, 51–52
Probability density functions, 29–32
  chi-square distribution, 32
  exponential distribution, 30–31
    algorithm, 30
  Gaussian processes, 30
    example, 30
      solution, 30
  lognormal distribution, 31–32
  normal distribution, 31
    algorithm, 31
  white noise, 29
Projection operator, 184–185
  example, 185

R

Random signals, distributions, 19–22
  book MATLAB script file, 20–21
  ergodic processes, 21–22
  stationary processes, 21–22
Random variables
  sequences, 19–54
    averages, 22–26
      correlation, 22–26
        example, 23
          solution, 23
      covariance, 25–26
      independent random variables, 26
      mean value, 22
      uncorrelated random variables, 26
    filtering random processes, 34–36
      example, 35
        solution, 35
      spectral factorization, 35–36
    nonparametric spectra estimation, 40–49
      Bartlett method, 44
      Blackman–Tukey method, 42–44
        Bartlett (triangle) window, 43
        Blackman window, 43
        Hamming window, 43
        Hann window, 43
        Kaiser window, 43
        rectangle window, 42
      Blackman–Tukey periodogram with Bartlett window, 48
      correlogram, 40
      modified Welch method, 46

periodogram, 40
Welch method, 45–46
parametric methods of power spectral estimations, 49–54
autoregressive process, 50–51
hints-solutions-suggestions, 52–54
moving average process, 50
problems, 51–52
random signals, distributions, 19–22
book MATLAB script file, 20–21
ergodic processes, 21–22
stationary processes, 21–22
special random signals and probability density functions, 29–32
chi-square distribution, 32
exponential distribution, 30–31
algorithm, 30
Gaussian processes, 30
example, 30
solution, 30
lognormal distribution, 31–32
normal distribution, 31
algorithm, 31
white noise, 29
special types of random processes, 36–40
autoregressive moving average process, 36–38
autoregressive process, 38
example, 38
solution, 38–39
moving average process, 40
stationary processes, 26–29
autocorrelation matrix, 27
example, 27–29
solution, 27–29
Wiener–Khintchin relations, 32–34
example, 33
solution, 33
signals, 19–54
autoregressive moving average process, 36–38
autoregressive process, 38
example, 38
solution, 38–39
averages, 22–26
correlation, 22–26
example, 23
solution, 23
covariance, 25–26
independent random variables, 26
mean value, 22
uncorrelated random variables, 26
filtering random processes, 34–36
example, 35
solution, 35
spectral factorization, 35–36
moving average process, 40
nonparametric spectra estimation, 40–49
Bartlett method, 44
Blackman–Tukey method, 42–44
Bartlett (triangle) window, 43
Blackman window, 43
Hamming window, 43
Hann window, 43
Kaiser window, 43
rectangle window, 42
Blackman–Tukey periodogram with Bartlett window, 48
correlogram, 40
modified Welch method, 46
periodogram, 40
Welch method, 45–46
parametric methods, power spectral estimations, 49–54
autoregressive process, 50–51
hints-solutions-suggestions, 52–54
moving average process, 50
problems, 51–52
probability density functions, 29–32
chi-square distribution, 32
exponential distribution, 30–31
algorithm, 30
Gaussian processes, 30
example, 30
solution, 30
lognormal distribution, 31–32
normal distribution, 31
algorithm, 31
white noise, 29
random signals, distributions, 19–22
book MATLAB script file, 20–21
ergodic processes, 21–22
stationary processes, 21–22
stationary processes, 26–29
autocorrelation matrix, 27
example, 27–29
solution, 27–29
Wiener–Khintchin relations, 32–34
example, 33
solution, 33
Recursive least-square algorithm, 188–201
example, 194
hints-solutions-suggestions, 197–201
initialization of, 193–194
initialization of recursive least-square algorithm, 193–194
matrix inversion lemma, 191

problems, 197–201
summary, 194–196
RLS. *See* Recursive least-square algorithm

S

Self-correcting adaptive filtering, 150–153
Sign algorithms, 137–139
error sign algorithm, 137–138
normalized least mean-square sign algorithm, 138
sign-sign algorithm, 139
signed-regressor algorithm, 139
Special random signals and probability density functions, 29–32
chi-square distribution, 32
exponential distribution, 30–31
algorithm, 30
Gaussian processes, 30
example, 30
solution, 30
lognormal distribution, 31–32
normal distribution, 31
algorithm, 31
white noise, 29
Special types of random processes, 36–40
autoregressive moving average process, 36–38
autoregressive process, 38
example, 38
solution, 38–39
moving average process, 40
Spectral factorization, 35–36
filtering random processes, 35–36
Stationary processes, 21–22, 26–29
autocorrelation matrix, 27
example, 27–29
solution, 27–29
Steepest-descent algorithm, 91–99
hints-solutions-suggestions, 97–99
problems, 96–97
solution of vector difference equation, 94–96
example, 96
stability (convergence) of algorithm, 92–94
transient behavior of mean-square error, 94

T

Transform domain adaptive least mean-square filtering, 153–157
convergence, 156–157
example, 155–156
Transform-domain representation, discrete-time signals, 6–11
discrete Fourier transform, 8
example, 8–9
solution, 9–10
discrete-time Fourier transform, 6–11
example, 6–7
solution, 7

V

Variable step-size least mean-square algorithm, 141–142

W

White noise, 29
Wiener filters, 55–76
book MATLAB function for noise canceling, 71
self-correcting Wiener filter example, 72
example, 63–67
finite impulse response Wiener filter, 55–59
example, 57
solution, 57–58
hints-solutions-suggestions, 74–76
mean-square error, 55
minimum mean-square error, 67
optimum filter, 67–71
noise canceling example, 69–71
system identification example, 68–69
problems, 73–74
Wiener filtering examples, 63–76
book MATLAB function for noise canceling, 71
self-correcting Wiener filter example, 72
example, 63–67
hints-solutions-suggestions, 74–76
minimum mean-square error, 67
optimum filter, 67–71
noise canceling example, 69–71
system identification example, 68–69
problems, 73–74
Wiener solution, 59–62
example, 60–61
solution, 60
orthogonality condition, 62
Wiener–Khintchin relations, 32–34
example, 33
solution, 33

Z

Z-transform, 11–12
example, 11
solution, 11–12